THE COURT MIDWIFE

A Series Edited by Margaret L. King and Albert Rabil Jr.

RECENT BOOKS IN THE SERIES

MARIA GAETANA AGNESI ET ALIA
The Contest for Knowledge: Debates over Women's Learning in Eighteenth-Century Italy
Edited and Translated by Rebecca Messbarger and Paula Findlen
Introduction by Rebecca Messbarger

FRANCISCA DE LOS APÓSTOLES
The Inquisition of Francisca: A Sixteenth-Century Visionary on Trial
Edited and Translated by Gillian T. W. Ahlgren

GIULIA BIGOLINA
Urania: A Romance
Edited and Translated by Valeria Finucci

MADDALENA CAMPIGLIA
Flori, a Pastoral Drama: A Bilingual Edition
Edited and with an Introduction and Notes by Virginia Cox and Lisa Sampson
Translated by Virginia Cox

VITTORIA COLONNA
Sonnets for Michelangelo: A Bilingual Edition
Edited and Translated by Abigail Brundin

MARIE DENTIÈRE
Epistle to Marguerite de Navarre and Preface to a Sermon by John Calvin
Edited and Translated by Mary B. McKinley

MADAME DE MAINTENON
Dialogues and Addresses
Edited and Translated by John J. Conley, S.J.

ISOTTA NOGAROLA
Complete Writings: Letterbook, Dialogue on Adam and Eve, Orations
Edited and Translated by Margaret L. King and Diana Robin

JOHANNA ELEONORA PETERSEN
The Life of Lady Johanna Eleonora Petersen, Written by Herself: Pietism and Women's Autobiography
Edited and Translated by Barbara Becker-Cantarino

MADELEINE DE SCUDÉRY
Selected Letters, Orations, and Rhetorical Dialogues
Edited and Translated by Jane Donawerth and Julie Strongson

ELISABETTA CAMINER TURRA
Selected Writings of an Eighteenth-Century Venetian Woman of Letters
Edited and Translated by Catherine M. Sama

MADAME DE VILLEDIEU
(MARIE-CATHERINE DESJARDINS)
Memoirs of the Life of Henriette-Sylvie de Molière: A Novel
Edited and Translated by Donna Kuizenga

Justine Siegemund

THE COURT MIDWIFE

Edited and Translated by Lynne Tatlock

THE UNIVERSITY OF CHICAGO PRESS
Chicago & London

Justine Siegemund, 1636–1706

Lynne Tatlock is the Hortense and Tobias Lewin Distinguished Professor in Humanities in the Department of Germanic Languages and Literatures at Washington University in St. Louis. She is the author of a monograph on Willibald Alexis (1984), the translator of two books, most recently of Gabriele Reuter's *From a Good Family* (1999), and the editor or coeditor of six books, most recently of *German Culture in Nineteenth-Century America: Reception, Adaptation, and Transformation* (2005).

The University of Chicago Press, Chicago 60637
The University of Chicago Press, Ltd., London

Printed in the United States of America

14 13 12 11 10 09 08 07 06 05 1 2 3 4 5

ISBN: 0-226-75708-0 (cloth)
ISBN: 0-226-75709-9 (paper)

The illustrations in this book are reproduced from the 1690 edition of Justine Siegemund's *Die chur-brandenburgische hoff-wehe-mutter* by permission of the University of Chicago Library, Special Collections Research Center.

Library of Congress Cataloging-in-Publication Data

Siegemund, Justina, 1636–1706.
The court midwife / Justina Siegemund ; edited and translated by Lynne Tatlock.
p. cm.—(The other voice in early modern Europe)
Includes bibliographical references and index.
ISBN 0-226-75708-0 (cloth : alk. paper)—ISBN 0-226-75709-9 (pbk. : alk. paper)
1. Midwifery—Early works to 1800. I. Tatlock, Lynne, 1950–
II. Title. III. Series.
RG950.S54 2005
618.2'009—dc22
2005002438

♾ The paper used in this publication meets the minimum requirements of the American National Standard for Information Sciences—Permanence of Paper for Printed Library Materials, ANSI Z39.48-1992.

For Renate

CONTENTS

ACKNOWLEDGMENTS

Alluding in 1671 to her lack of knowledge of Greek, the language of academic medicine, the English midwife Jane Sharp maintained, "It is not hard Words that perform the Work, as if none understood the Art that cannot understand Greek. Words are but the Shell, that we oftentimes break our Teeth with them to come at the Kernel, I mean our Brains, to know what is the Meaning of them . . . ." During the protracted work of translating *The Court Midwife,* I have frequently felt as if I were breaking my brain in trying to "come at the kernel" of Justine Siegemund's seventeenth-century German words. I had, however, the good fortune of having the support and advice of friends and colleagues, as well as access to excellent libraries.

I am grateful to Gerhild Scholz Williams for cheerfully answering questions about murky passages and to Michael Sherberg and George Pepe for help with Latin phrases. Gwyneth Cliver, Alyssa Lonner, and especially Erika Rinker provided invaluable assistance by painstakingly comparing sections of my translation in progress with the original German and checking for and detecting oversights. Alyssa Lonner kindly tracked down English equivalents of German proverbs. I thank Otti Reichl, retired midwife in Germany, for answering questions about her work and providing insight into Justine Siegemund's text. Walton Schalick, assistant professor of pediatrics and assistant professor of history, deserves special appreciation for bridging the gap between Arts and Sciences and the Medical School at Washington University and seeing to it that I eventually was able to observe live procedures. Amy Ravin, M.D., Grace Hill Neighborhood Health Centers Inc. and Clinical Instructor in the Washington University Department of Obstetrics and Gynecology, not only generously examined obscure passages of the text with me and recommended videos and medical texts but also saw to it that I observed a variety of gynecological and obstetric procedures and, most important, provided me with the priceless opportunity to observe several live births.

I thank the Universtätsbibliothek Göttingen, Germany; the Herzog August Bibliothek in Wolfenbüttel, Germany; the Medical Library of Washington University in St. Louis; and the Regenstein Library of the University of Chicago for providing me access to their rich collections of early modern books on obstetrics and gynecology and to the various editions of Siegemund's book that they hold. I am particularly grateful to Barbara Gilbert of the University of Chicago and Elaine Challacombe of the Bio-Medical Library of the University of Minnesota, who both responded promptly to my inquiries. As a result, I discovered the existence of three previously unrecorded dedications, including that by Mary II of England, dedications that brought into focus the ambition of Siegemund's undertaking. I am also indebted to rare book librarian Lilla Vekerdy of Washington University's Medical School who in this project, as always, has ever been ready to do a little more than I asked for.

I am grateful to the editors, Margaret L. King and Albert Rabil, and the University of Chicago Press for including Siegemund in the Other Voice series. A collaborative research grant from the National Endowment for the Humanities that was awarded to several projects in the series helped finance the cost of travel to Germany, as well as research and other expenses involved in producing the present volume. I thank the editorial staff at the press, especially Randy Petilos, who shepherded the project through its initial phase at the press, and Maia Rigas, who valiantly saw it through to completion. Arts and Sciences at Washington University in St. Louis generously helped to defray some of the considerable production costs generated by a volume with over forty illustrations.

My husband, Joseph F. Loewenstein, deserves thanks for bearing up under a yearlong barrage of questions about English usage and of dinnertime reporting on Justine Siegemund and for his unfailingly generous and useful advice. Finally, I am deeply indebted to my friend Renate Schmidt, whom I have known since my student days in Germany and who obligingly responded from Germany by e-mail for months on end to my questions about the meaning of Siegemund's words. I can neither imagine having achieved fluency in German in the first place without her nor could I feel half so confident about the accuracy of the present translation; in gratitude for her German words and many years of friendship I dedicate this work to her.

Lynne Tatlock

THE OTHER VOICE IN EARLY MODERN EUROPE: INTRODUCTION TO THE SERIES

Margaret L. King and Albert Rabil Jr.

THE OLD VOICE AND THE OTHER VOICE

In western Europe and the United States, women are nearing equality in the professions, in business, and in politics. Most enjoy access to education, reproductive rights, and autonomy in financial affairs. Issues vital to women are on the public agenda: equal pay, child care, domestic abuse, breast cancer research, and curricular revision with an eye to the inclusion of women.

These recent achievements have their origins in things women (and some male supporters) said for the first time about six hundred years ago. Theirs is the "other voice," in contradistinction to the "first voice," the voice of the educated men who created Western culture. Coincident with a general reshaping of European culture in the period 1300–1700 (called the Renaissance or early modern period), questions of female equality and opportunity were raised that still resound and are still unresolved.

The other voice emerged against the backdrop of a three-thousand-year history of the derogation of women rooted in the civilizations related to Western culture: Hebrew, Greek, Roman, and Christian. Negative attitudes toward women inherited from these traditions pervaded the intellectual, medical, legal, religious, and social systems that developed during the European Middle Ages.

The following pages describe the traditional, overwhelmingly male views of women's nature inherited by early modern Europeans and the new tradition that the "other voice" called into being to begin to challenge reigning assumptions. This review should serve as a framework for understanding the texts published in the series the Other Voice in Early Modern Europe. Introductions specific to each text and author follow this essay in all the volumes of the series.

TRADITIONAL VIEWS OF WOMEN, 500 B.C.E.–1500 C.E.

Embedded in the philosophical and medical theories of the ancient Greeks were perceptions of the female as inferior to the male in both mind and body. Similarly, the structure of civil legislation inherited from the ancient Romans was biased against women, and the views on women developed by Christian thinkers out of the Hebrew Bible and the Christian New Testament were negative and disabling. Literary works composed in the vernacular of ordinary people, and widely recited or read, conveyed these negative assumptions. The social networks within which most women lived—those of the family and the institutions of the Roman Catholic Church—were shaped by this negative tradition and sharply limited the areas in which women might act in and upon the world.

GREEK PHILOSOPHY AND FEMALE NATURE. Greek biology assumed that women were inferior to men and defined them as merely childbearers and housekeepers. This view was authoritatively expressed in the works of the philosopher Aristotle.

Aristotle thought in dualities. He considered action superior to inaction, form (the inner design or structure of any object) superior to matter, completion to incompletion, possession to deprivation. In each of these dualities, he associated the male principle with the superior quality and the female with the inferior. "The male principle in nature," he argued, "is associated with active, formative and perfected characteristics, while the female is passive, material and deprived, desiring the male in order to become complete."[1] Men are always identified with virile qualities, such as judgment, courage, and stamina, and women with their opposites—irrationality, cowardice, and weakness.

The masculine principle was considered superior even in the womb. The man's semen, Aristotle believed, created the form of a new human creature, while the female body contributed only matter. (The existence of the ovum, and with it the other facts of human embryology, was not established until the seventeenth century.) Although the later Greek physician Galen believed there was a female component in generation, contributed by "female semen," the followers of both Aristotle and Galen saw the male role in human generation as more active and more important.

In the Aristotelian view, the male principle sought always to reproduce

1. Aristotle, *Physics* 1.9.192a20–24, in *The Complete Works of Aristotle,* ed. Jonathan Barnes, rev. Oxford trans., 2 vols. (Princeton, 1984), 1:328.

itself. The creation of a female was always a mistake, therefore, resulting from an imperfect act of generation. Every female born was considered a "defective" or "mutilated" male (as Aristotle's terminology has variously been translated), a "monstrosity" of nature.[2]

For Greek theorists, the biology of males and females was the key to their psychology. The female was softer and more docile, more apt to be despondent, querulous, and deceitful. Being incomplete, moreover, she craved sexual fulfillment in intercourse with a male. The male was intellectual, active, and in control of his passions.

These psychological polarities derived from the theory that the universe consisted of four elements (earth, fire, air, and water), expressed in human bodies as four "humors" (black bile, yellow bile, blood, and phlegm) considered, respectively, dry, hot, damp, and cold and corresponding to mental states ("melancholic," "choleric," "sanguine," "phlegmatic"). In this scheme the male, sharing the principles of earth and fire, was dry and hot; the female, sharing the principles of air and water, was cold and damp.

Female psychology was further affected by her dominant organ, the uterus (womb), *hystera* in Greek. The passions generated by the womb made women lustful, deceitful, talkative, irrational, indeed—when these affects were in excess—"hysterical."

Aristotle's biology also had social and political consequences. If the male principle was superior and the female inferior, then in the household, as in the state, men should rule and women must be subordinate. That hierarchy did not rule out the companionship of husband and wife, whose cooperation was necessary for the welfare of children and the preservation of property. Such mutuality supported male preeminence.

Aristotle's teacher Plato suggested a different possibility: that men and women might possess the same virtues. The setting for this proposal is the imaginary and ideal Republic that Plato sketches in a dialogue of that name. Here, for a privileged elite capable of leading wisely, all distinctions of class and wealth dissolve, as, consequently, do those of gender. Without households or property, as Plato constructs his ideal society, there is no need for the subordination of women. Women may therefore be educated to the same level as men to assume leadership. Plato's Republic remained imaginary, however. In real societies, the subordination of women remained the norm and the prescription.

The views of women inherited from the Greek philosophical tradition became the basis for medieval thought. In the thirteenth century, the su-

2. Aristotle, *Generation of Animals* 2.3.737a27–28, in *The Complete Works,* 1:1144.

preme Scholastic philosopher Thomas Aquinas, among others, still echoed Aristotle's views of human reproduction, of male and female personalities, and of the preeminent male role in the social hierarchy.

ROMAN LAW AND THE FEMALE CONDITION. Roman law, like Greek philosophy, underlay medieval thought and shaped medieval society. The ancient belief that adult property-owning men should administer households and make decisions affecting the community at large is the very fulcrum of Roman law.

About 450 B.C.E., during Rome's republican era, the community's customary law was recorded (legendarily) on twelve tablets erected in the city's central forum. It was later elaborated by professional jurists whose activity increased in the imperial era, when much new legislation was passed, especially on issues affecting family and inheritance. This growing, changing body of laws was eventually codified in the *Corpus of Civil Law* under the direction of the emperor Justinian, generations after the empire ceased to be ruled from Rome. That *Corpus,* read and commented on by medieval scholars from the eleventh century on, inspired the legal systems of most of the cities and kingdoms of Europe.

Laws regarding dowries, divorce, and inheritance pertain primarily to women. Since those laws aimed to maintain and preserve property, the women concerned were those from the property-owning minority. Their subordination to male family members points to the even greater subordination of lower-class and slave women, about whom the laws speak little.

In the early republic, the *paterfamilias,* or "father of the family," possessed *patria potestas,* "paternal power." The term *pater,* "father," in both these cases does not necessarily mean biological father but denotes the head of a household. The father was the person who owned the household's property and, indeed, its human members. The *paterfamilias* had absolute power—including the power, rarely exercised, of life or death—over his wife, his children, and his slaves, as much as his cattle.

Male children could be "emancipated," an act that granted legal autonomy and the right to own property. Those over fourteen could be emancipated by a special grant from the father or automatically by their father's death. But females could never be emancipated; instead, they passed from the authority of their father to that of a husband or, if widowed or orphaned while still unmarried, to a guardian or tutor.

Marriage in its traditional form placed the woman under her husband's authority, or *manus.* He could divorce her on grounds of adultery, drinking wine, or stealing from the household, but she could not divorce him. She could neither possess property in her own right nor bequeath any to her children upon her death. When her husband died, the household property passed not to her but to his male heirs. And when her father died, she had no claim to any family inheritance, which was directed to her brothers or more remote male relatives. The effect of these laws was to exclude women from civil society, itself based on property ownership.

In the later republican and imperial periods, these rules were significantly modified. Women rarely married according to the traditional form. The practice of "free" marriage allowed a woman to remain under her father's authority, to possess property given her by her father (most frequently the "dowry," recoverable from the husband's household on his death), and to inherit from her father. She could also bequeath property to her own children and divorce her husband, just as he could divorce her.

Despite this greater freedom, women still suffered enormous disability under Roman law. Heirs could belong only to the father's side, never the mother's. Moreover, although she could bequeath her property to her children, she could not establish a line of succession in doing so. A woman was "the beginning and end of her own family," said the jurist Ulpian. Moreover, women could play no public role. They could not hold public office, represent anyone in a legal case, or even witness a will. Women had only a private existence and no public personality.

The dowry system, the guardian, women's limited ability to transmit wealth, and total political disability are all features of Roman law adopted by the medieval communities of western Europe, although modified according to local customary laws.

CHRISTIAN DOCTRINE AND WOMEN'S PLACE. The Hebrew Bible and the Christian New Testament authorized later writers to limit women to the realm of the family and to burden them with the guilt of original sin. The passages most fruitful for this purpose were the creation narratives in Genesis and sentences from the Epistles defining women's role within the Christian family and community.

Each of the first two chapters of Genesis contains a creation narrative. In the first "God created man in his own image, in the image of God he created him; male and female he created them" (Gn 1:27). In the second, God created Eve from Adam's rib (2:21–23). Christian theologians relied principally on Genesis 2 for their understanding of the relation between man and woman, interpreting the creation of Eve from Adam as proof of her subordination to him.

The creation story in Genesis 2 leads to that of the temptations in Genesis 3: of Eve by the wily serpent and of Adam by Eve. As read by Christian theologians from Tertullian to Thomas Aquinas, the narrative made Eve responsible for the Fall and its consequences. She instigated the act; she deceived her husband; she suffered the greater punishment. Her disobedience made it necessary for Jesus to be incarnated and to die on the cross. From the pulpit, moralists and preachers for centuries conveyed to women the guilt that they bore for original sin.

The Epistles offered advice to early Christians on building communities of the faithful. Among the matters to be regulated was the place of women. Paul offered views favorable to women in Galatians 3:28: "There is neither Jew nor Greek, there is neither slave nor free, there is neither male nor female; for you are all one in Christ Jesus." Paul also referred to women as his coworkers and placed them on a par with himself and his male coworkers (Phlm 4:2–3; Rom 16:1–3; 1 Cor 16:19). Elsewhere, Paul limited women's possibilities: "But I want you to understand that the head of every man is Christ, the head of a woman is her husband, and the head of Christ is God" (1 Cor 11:3).

Biblical passages by later writers (although attributed to Paul) enjoined women to forgo jewels, expensive clothes, and elaborate coiffures; and they forbade women to "teach or have authority over men," telling them to "learn in silence with all submissiveness" as is proper for one responsible for sin, consoling them, however, with the thought that they will be saved through childbearing (1 Tm 2:9–15). Other texts among the later Epistles defined women as the weaker sex and emphasized their subordination to their husbands (1 Pt 3:7; Col 3:18; Eph 5:22–23).

These passages from the New Testament became the arsenal employed by theologians of the early church to transmit negative attitudes toward women to medieval Christian culture—above all, Tertullian (*On the Apparel of Women*), Jerome (*Against Jovinian*), and Augustine (*The Literal Meaning of Genesis*).

THE IMAGE OF WOMEN IN MEDIEVAL LITERATURE. The philosophical, legal, and religious traditions born in antiquity formed the basis of the medieval intellectual synthesis wrought by trained thinkers, mostly clerics, writing in Latin and based largely in universities. The vernacular literary tradition that developed alongside the learned tradition also spoke about female nature and women's roles. Medieval stories, poems, and epics also portrayed women negatively—as lustful and deceitful—while praising good housekeepers and loyal wives as replicas of the Virgin Mary or the female saints and martyrs.

There is an exception in the movement of "courtly love" that evolved in southern France from the twelfth century. Courtly love was the erotic love between a nobleman and noblewoman, the latter usually superior in social rank. It was always adulterous. From the conventions of courtly love derive modern Western notions of romantic love. The tradition has had an impact disproportionate to its size, for it affected only a tiny elite, and very few women. The exaltation of the female lover probably does not reflect a higher evaluation of women or a step toward their sexual liberation. More likely it gives expression to the social and sexual tensions besetting the knightly class at a specific historical juncture.

The literary fashion of courtly love was on the wane by the thirteenth century, when the widely read *Romance of the Rose* was composed in French by two authors of significantly different dispositions. Guillaume de Lorris composed the initial four thousand verses about 1235, and Jean de Meun added about seventeen thousand verses—more than four times the original—about 1265.

The fragment composed by Guillaume de Lorris stands squarely in the tradition of courtly love. Here the poet, in a dream, is admitted into a walled garden where he finds a magic fountain in which a rosebush is reflected. He longs to pick one rose, but the thorns prevent his doing so, even as he is wounded by arrows from the god of love, whose commands he agrees to obey. The rest of this part of the poem recounts the poet's unsuccessful efforts to pluck the rose.

The longer part of the *Romance* by Jean de Meun also describes a dream. But here allegorical characters give long didactic speeches, providing a social satire on a variety of themes, some pertaining to women. Love is an anxious and tormented state, the poem explains: women are greedy and manipulative, marriage is miserable, beautiful women are lustful, ugly ones cease to please, and a chaste woman is as rare as a black swan.

Shortly after Jean de Meun completed *The Romance of the Rose,* Mathéolus penned his *Lamentations,* a long Latin diatribe against marriage translated into French about a century later. The *Lamentations* sum up medieval attitudes toward women and provoked the important response by Christine de Pizan in her *Book of the City of Ladies.*

In 1355, Giovanni Boccaccio wrote *Il Corbaccio,* another antifeminist manifesto, although ironically by an author whose other works pioneered new directions in Renaissance thought. The former husband of his lover appears to Boccaccio, condemning his unmoderated lust and detailing the defects of women. Boccaccio concedes at the end "how much men naturally surpass women in nobility" and is cured of his desires.[3]

3. Giovanni Boccaccio, *The Corbaccio, or The Labyrinth of Love,* trans. and ed. Anthony K. Cassell, rev. ed. (Binghamton, NY, 1993), 71.

WOMEN'S ROLES: THE FAMILY. The negative perceptions of women expressed in the intellectual tradition are also implicit in the actual roles that women played in European society. Assigned to subordinate positions in the household and the church, they were barred from significant participation in public life.

Medieval European households, like those in antiquity and in non-Western civilizations, were headed by males. It was the male serf (or peasant), feudal lord, town merchant, or citizen who was polled or taxed or succeeded to an inheritance or had any acknowledged public role, although his wife or widow could stand as a temporary surrogate. From about 1100, the position of property-holding males was further enhanced: inheritance was confined to the male, or agnate, line—with depressing consequences for women.

A wife never fully belonged to her husband's family, nor was she a daughter to her father's family. She left her father's house young to marry whomever her parents chose. Her dowry was managed by her husband, and at her death it normally passed to her children by him.

A married woman's life was occupied nearly constantly with cycles of pregnancy, childbearing, and lactation. Women bore children through all the years of their fertility, and many died in childbirth. They were also responsible for raising young children up to six or seven. In the propertied classes that responsibility was shared, since it was common for a wet nurse to take over breast-feeding and for servants to perform other chores.

Women trained their daughters in the household duties appropriate to their status, nearly always tasks associated with textiles: spinning, weaving, sewing, embroidering. Their sons were sent out of the house as apprentices or students, or their training was assumed by fathers in later childhood and adolescence. On the death of her husband, a woman's children became the responsibility of his family. She generally did not take "his" children with her to a new marriage or back to her father's house, except sometimes in the artisan classes.

Women also worked. Rural peasants performed farm chores, merchant wives often practiced their husbands' trades, the unmarried daughters of the urban poor worked as servants or prostitutes. All wives produced or embellished textiles and did the housekeeping, while wealthy ones managed servants. These labors were unpaid or poorly paid but often contributed substantially to family wealth.

WOMEN'S ROLES: THE CHURCH. Membership in a household, whether a father's or a husband's, meant for women a lifelong subordination to others. In western Europe, the Roman Catholic Church offered an alternative to the career of wife and mother. A woman could enter a convent, parallel in function to the monasteries for men that evolved in the early Christian centuries.

In the convent, a woman pledged herself to a celibate life, lived according to strict community rules, and worshiped daily. Often the convent offered training in Latin, allowing some women to become considerable scholars and authors as well as scribes, artists, and musicians. For women who chose the conventual life, the benefits could be enormous, but for numerous others placed in convents by paternal choice, the life could be restrictive and burdensome.

The conventual life declined as an alternative for women as the modern age approached. Reformed monastic institutions resisted responsibility for related female orders. The church increasingly restricted female institutional life by insisting on closer male supervision.

Women often sought other options. Some joined the communities of laywomen that sprang up spontaneously in the thirteenth century in the urban zones of western Europe, especially in Flanders and Italy. Some joined the heretical movements that flourished in late medieval Christendom, whose anticlerical and often antifamily positions particularly appealed to women. In these communities, some women were acclaimed as "holy women" or "saints," whereas others often were condemned as frauds or heretics.

In all, although the options offered to women by the church were sometimes less than satisfactory, they were sometimes richly rewarding. After 1520, the convent remained an option only in Roman Catholic territories. Protestantism engendered an ideal of marriage as a heroic endeavor and appeared to place husband and wife on a more equal footing. Sermons and treatises, however, still called for female subordination and obedience.

THE OTHER VOICE, 1300–1700

When the modern era opened, European culture was so firmly structured by a framework of negative attitudes toward women that to dismantle it was a monumental labor. The process began as part of a larger cultural movement that entailed the critical reexamination of ideas inherited from the ancient and medieval past. The humanists launched that critical reexamination.

THE HUMANIST FOUNDATION. Originating in Italy in the fourteenth century, humanism quickly became the dominant intellectual movement in Europe. Spreading in the sixteenth century from Italy to the rest of Europe, it fueled the literary, scientific, and philosophical movements of the era and laid the basis for the eighteenth-century Enlightenment.

Humanists regarded the Scholastic philosophy of medieval universities as out of touch with the realities of urban life. They found in the rhetorical discourse of classical Rome a language adapted to civic life and public speech. They learned to read, speak, and write classical Latin and, eventually, classical Greek. They founded schools to teach others to do so, establishing the pattern for elementary and secondary education for the next three hundred years.

In the service of complex government bureaucracies, humanists employed their skills to write eloquent letters, deliver public orations, and formulate public policy. They developed new scripts for copying manuscripts and used the new printing press to disseminate texts, for which they created methods of critical editing.

Humanism was a movement led by males who accepted the evaluation of women in ancient texts and generally shared the misogynist perceptions of their culture. (Female humanists, as we will see, did not.) Yet humanism also opened the door to a reevaluation of the nature and capacity of women. By calling authors, texts, and ideas into question, it made possible the fundamental rereading of the whole intellectual tradition that was required in order to free women from cultural prejudice and social subordination.

A DIFFERENT CITY. The other voice first appeared when, after so many centuries, the accumulation of misogynist concepts evoked a response from a capable female defender: Christine de Pizan (1365–1431). Introducing her *Book of the City of Ladies* (1405), she described how she was affected by reading Mathéolus's *Lamentations*: "Just the sight of this book . . . made me wonder how it happened that so many different men . . . are so inclined to express both in speaking and in their treatises and writings so many wicked insults about women and their behavior."[4] These statements impelled her to detest herself "and the entire feminine sex, as though we were monstrosities in nature."[5]

The rest of *The Book of the City of Ladies* presents a justification of the female sex and a vision of an ideal community of women. A pioneer, she has received the message of female inferiority and rejected it. From the fourteenth to the seventeenth century, a huge body of literature accumulated that responded to the dominant tradition.

4. Christine de Pizan, *The Book of the City of Ladies,* trans. Earl Jeffrey Richards, foreword by Marina Warner (New York, 1982), 1.1.1, pp. 3–4.

5. Ibid., 1.1.1–2, p. 5.

The result was a literary explosion consisting of works by both men and women, in Latin and in the vernaculars: works enumerating the achievements of notable women; works rebutting the main accusations made against women; works arguing for the equal education of men and women; works defining and redefining women's proper role in the family, at court, in public; works describing women's lives and experiences. Recent monographs and articles have begun to hint at the great range of this movement, involving probably several thousand titles. The protofeminism of these "other voices" constitutes a significant fraction of the literary product of the early modern era.

THE CATALOGS. About 1365, the same Boccaccio whose *Corbaccio* rehearses the usual charges against female nature wrote another work, *Concerning Famous Women*. A humanist treatise drawing on classical texts, it praised 106 notable women: ninety-eight of them from pagan Greek and Roman antiquity, one (Eve) from the Bible, and seven from the medieval religious and cultural tradition; his book helped make all readers aware of a sex normally condemned or forgotten. Boccaccio's outlook nevertheless was unfriendly to women, for it singled out for praise those women who possessed the traditional virtues of chastity, silence, and obedience. Women who were active in the public realm—for example, rulers and warriors—were depicted as usually being lascivious and as suffering terrible punishments for entering the masculine sphere. Women were his subject, but Boccaccio's standard remained male.

Christine de Pizan's *Book of the City of Ladies* contains a second catalog, one responding specifically to Boccaccio's. Whereas Boccaccio portrays female virtue as exceptional, she depicts it as universal. Many women in history were leaders, or remained chaste despite the lascivious approaches of men, or were visionaries and brave martyrs.

The work of Boccaccio inspired a series of catalogs of illustrious women of the biblical, classical, Christian, and local pasts, among them Filippo da Bergamo's *Of Illustrious Women*, Pierre de Brantôme's *Lives of Illustrious Women*, Pierre Le Moyne's *Gallerie of Heroic Women*, and Pietro Paolo de Ribera's *Immortal Triumphs and Heroic Enterprises of 845 Women*. Whatever their embedded prejudices, these works drove home to the public the possibility of female excellence.

THE DEBATE. At the same time, many questions remained: Could a woman be virtuous? Could she perform noteworthy deeds? Was she even, strictly speaking, of the same human species as men? These questions were debated over four centuries, in French, German, Italian, Spanish, and English, by authors male and female, among Catholics, Protestants, and Jews, in ponderous volumes and breezy pamphlets. The whole literary genre has been called the *querelle des femmes,* the "woman question."

The opening volley of this battle occurred in the first years of the fifteenth century, in a literary debate sparked by Christine de Pizan. She exchanged letters critical of Jean de Meun's contribution to *The Romance of the Rose* with two French royal secretaries, Jean de Montreuil and Gontier Col. When the matter became public, Jean Gerson, one of Europe's leading theologians, supported de Pizan's arguments against de Meun, for the moment silencing the opposition.

The debate resurfaced repeatedly over the next two hundred years. *The Triumph of Women* (1438) by Juan Rodríguez de la Camara (or Juan Rodríguez del Padron) struck a new note by presenting arguments for the superiority of women to men. *The Champion of Women* (1440–42) by Martin Le Franc addresses once again the negative views of women presented in *The Romance of the Rose* and offers counterevidence of female virtue and achievement.

A cameo of the debate on women is included in *The Courtier,* one of the most widely read books of the era, published by the Italian Baldassare Castiglione in 1528 and immediately translated into other European vernaculars. *The Courtier* depicts a series of evenings at the court of the duke of Urbino in which many men and some women of the highest social stratum amuse themselves by discussing a range of literary and social issues. The "woman question" is a pervasive theme throughout, and the third of its four books is devoted entirely to that issue.

In a verbal duel, Gasparo Pallavicino and Giuliano de' Medici present the main claims of the two traditions. Gasparo argues the innate inferiority of women and their inclination to vice. Only in bearing children do they profit the world. Giuliano counters that women share the same spiritual and mental capacities as men and may excel in wisdom and action. Men and women are of the same essence: just as no stone can be more perfectly a stone than another, so no human being can be more perfectly human than others, whether male or female. It was an astonishing assertion, boldly made to an audience as large as all Europe.

THE TREATISES. Humanism provided the materials for a positive counterconcept to the misogyny embedded in Scholastic philosophy and law and inherited from the Greek, Roman, and Christian pasts. A series of humanist treatises on marriage and family, on education and deportment, and on the nature of women helped construct these new perspectives.

The works by Francesco Barbaro and Leon Battista Alberti—*On Mar-*

riage (1415) and *On the Family* (1434–37)—far from defending female equality, reasserted women's responsibility for rearing children and managing the housekeeping while being obedient, chaste, and silent. Nevertheless, they served the cause of reexamining the issue of women's nature by placing domestic issues at the center of scholarly concern and reopening the pertinent classical texts. In addition, Barbaro emphasized the companionate nature of marriage and the importance of a wife's spiritual and mental qualities for the well-being of the family.

These themes reappear in later humanist works on marriage and the education of women by Juan Luis Vives and Erasmus. Both were moderately sympathetic to the condition of women without reaching beyond the usual masculine prescriptions for female behavior.

An outlook more favorable to women characterizes the nearly unknown work *In Praise of Women* (ca. 1487) by the Italian humanist Bartolommeo Goggio. In addition to providing a catalog of illustrious women, Goggio argued that male and female are the same in essence, but that women (reworking the Adam and Eve narrative from quite a new angle) are actually superior. In the same vein, the Italian humanist Mario Equicola asserted the spiritual equality of men and women in *On Women* (1501). In 1525, Galeazzo Flavio Capra (or Capella) published his work *On the Excellence and Dignity of Women.* This humanist tradition of treatises defending the worthiness of women culminates in the work of Henricus Cornelius Agrippa *On the Nobility and Preeminence of the Female Sex.* No work by a male humanist more succinctly or explicitly presents the case for female dignity.

THE WITCH BOOKS. While humanists grappled with the issues pertaining to women and family, other learned men turned their attention to what they perceived as a very great problem: witches. Witch-hunting manuals, explorations of the witch phenomenon, and even defenses of witches are not at first glance pertinent to the tradition of the other voice. But they do relate in this way: most accused witches were women. The hostility aroused by supposed witch activity is comparable to the hostility aroused by women. The evil deeds the victims of the hunt were charged with were exaggerations of the vices to which, many believed, all women were prone.

The connection between the witch accusation and the hatred of women is explicit in the notorious witch-hunting manual *The Hammer of Witches* (1486) by two Dominican inquisitors, Heinrich Krämer and Jacob Sprenger. Here the inconstancy, deceitfulness, and lustfulness traditionally associated with women are depicted in exaggerated form as the core features of witch behavior. These traits inclined women to make a bargain with the devil—

sealed by sexual intercourse—by which they acquired unholy powers. Such bizarre claims, far from being rejected by rational men, were broadcast by intellectuals. The German Ulrich Molitur, the Frenchman Nicolas Rémy, and the Italian Stefano Guazzo all coolly informed the public of sinister orgies and midnight pacts with the devil. The celebrated French jurist, historian, and political philosopher Jean Bodin argued that because women were especially prone to diabolism, regular legal procedures could properly be suspended in order to try those accused of this "exceptional crime."

A few experts such as the physician Johann Weyer, a student of Agrippa's, raised their voices in protest. In 1563, he explained the witch phenomenon thus, without discarding belief in diabolism: the devil deluded foolish old women afflicted by melancholia, causing them to believe they had magical powers. Weyer's rational skepticism, which had good credibility in the community of the learned, worked to revise the conventional views of women and witchcraft.

WOMEN'S WORKS. To the many categories of works produced on the question of women's worth must be added nearly all works written by women. A woman writing was in herself a statement of women's claim to dignity.

Only a few women wrote anything before the dawn of the modern era, for three reasons. First, they rarely received the education that would enable them to write. Second, they were not admitted to the public roles—as administrator, bureaucrat, lawyer or notary, or university professor—in which they might gain knowledge of the kinds of things the literate public thought worth writing about. Third, the culture imposed silence on women, considering speaking out a form of unchastity. Given these conditions, it is remarkable that any women wrote. Those who did before the fourteenth century were almost always nuns or religious women whose isolation made their pronouncements more acceptable.

From the fourteenth century on, the volume of women's writings rose. Women continued to write devotional literature, although not always as cloistered nuns. They also wrote diaries, often intended as keepsakes for their children; books of advice to their sons and daughters; letters to family members and friends; and family memoirs, in a few cases elaborate enough to be considered histories.

A few women wrote works directly concerning the "woman question," and some of these, such as the humanists Isotta Nogarola, Cassandra Fedele, Laura Cereta, and Olympia Morata, were highly trained. A few were professional writers, living by the income of their pens; the very first among them was Christine de Pizan, noteworthy in this context as in so many others. In addition to *The Book of the City of Ladies* and her critiques of *The Romance of the Rose,* she wrote *The Treasure of the City of Ladies* (a guide to social decorum for women), an advice book for her son, much courtly verse, and a full-scale history of the reign of King Charles V of France.

WOMEN PATRONS. Women who did not themselves write but encouraged others to do so boosted the development of an alternative tradition. Highly placed women patrons supported authors, artists, musicians, poets, and learned men. Such patrons, drawn mostly from the Italian elites and the courts of northern Europe, figure disproportionately as the dedicatees of the important works of early feminism.

For a start, it might be noted that the catalogs of Boccaccio and Alvaro de Luna were dedicated to the Florentine noblewoman Andrea Acciaiuoli and to Doña María, first wife of King Juan II of Castile, while the French translation of Boccaccio's work was commissioned by Anne of Brittany, wife of King Charles VIII of France. The humanist treatises of Goggio, Equicola, Vives, and Agrippa were dedicated, respectively, to Eleanora of Aragon, wife of Ercole I d'Este, Duke of Ferrara; to Margherita Cantelma of Mantua; to Catherine of Aragon, wife of King Henry VIII of England; and to Margaret, Duchess of Austria and regent of the Netherlands. As late as 1696, Mary Astell's *Serious Proposal to the Ladies, for the Advancement of Their True and Greatest Interest* was dedicated to Princess Anne of Denmark.

These authors presumed that their efforts would be welcome to female patrons, or they may have written at the bidding of those patrons. Silent themselves, perhaps even unresponsive, these loftily placed women helped shape the tradition of the other voice.

THE ISSUES. The literary forms and patterns in which the tradition of the other voice presented itself have now been sketched. It remains to highlight the major issues around which this tradition crystallizes. In brief, there are four problems to which our authors return again and again, in plays and catalogs, in verse and letters, in treatises and dialogues, in every language: the problem of chastity, the problem of power, the problem of speech, and the problem of knowledge. Of these the greatest, preconditioning the others, is the problem of chastity.

THE PROBLEM OF CHASTITY. In traditional European culture, as in those of antiquity and others around the globe, chastity was perceived as woman's quintessential virtue—in contrast to courage, or generosity, or leadership, or rationality, seen as virtues characteristic of men. Opponents of women charged them with insatiable lust. Women themselves and their defenders—without disputing the validity of the standard—responded that women were capable of chastity.

The requirement of chastity kept women at home, silenced them, isolated them, left them in ignorance. It was the source of all other impediments. Why was it so important to the society of men, of whom chastity was not required, and who more often than not considered it their right to violate the chastity of any woman they encountered?

Female chastity ensured the continuity of the male-headed household. If a man's wife was not chaste, he could not be sure of the legitimacy of his offspring. If they were not his and they acquired his property, it was not his household, but some other man's, that had endured. If his daughter was not chaste, she could not be transferred to another man's household as his wife, and he was dishonored.

The whole system of the integrity of the household and the transmission of property was bound up in female chastity. Such a requirement pertained only to property-owning classes, of course. Poor women could not expect to maintain their chastity, least of all if they were in contact with high-status men to whom all women but those of their own household were prey.

In Catholic Europe, the requirement of chastity was further buttressed by moral and religious imperatives. Original sin was inextricably linked with the sexual act. Virginity was seen as heroic virtue, far more impressive than, say, the avoidance of idleness or greed. Monasticism, the cultural institution that dominated medieval Europe for centuries, was grounded in the renunciation of the flesh. The Catholic reform of the eleventh century imposed a similar standard on all the clergy and a heightened awareness of sexual requirements on all the laity. Although men were asked to be chaste, female unchastity was much worse: it led to the devil, as Eve had led mankind to sin.

To such requirements, women and their defenders protested their innocence. Furthermore, following the example of holy women who had escaped the requirements of family and sought the religious life, some women began to conceive of female communities as alternatives both to family and to the cloister. Christine de Pizan's city of ladies was such a community. Moderata Fonte and Mary Astell envisioned others. The luxurious salons of the French *précieuses* of the seventeenth century, or the comfortable English drawing rooms of the next, may have been born of the same impulse. Here women not only might escape, if briefly, the subordinate position that life in the family entailed but might also make claims to power, exercise their capacity for speech, and display their knowledge.

THE PROBLEM OF POWER. Women were excluded from power: the whole cultural tradition insisted on it. Only men were citizens, only men bore arms, only men could be chiefs or lords or kings. There were exceptions that did not disprove the rule, when wives or widows or mothers took the place of men, awaiting their return or the maturation of a male heir. A woman who attempted to rule in her own right was perceived as an anomaly, a monster, at once a deformed woman and an insufficient male, sexually confused and consequently unsafe.

The association of such images with women who held or sought power explains some otherwise odd features of early modern culture. Queen Elizabeth I of England, one of the few women to hold full regal authority in European history, played with such male/female images—positive ones, of course—in representing herself to her subjects. She was a prince, and manly, even though she was female. She was also (she claimed) virginal, a condition absolutely essential if she was to avoid the attacks of her opponents. Catherine de' Medici, who ruled France as widow and regent for her sons, also adopted such imagery in defining her position. She chose as one symbol the figure of Artemisia, an androgynous ancient warrior-heroine who combined a female persona with masculine powers.

Power in a woman, without such sexual imagery, seems to have been indigestible by the culture. A rare note was struck by the Englishman Sir Thomas Elyot in his *Defence of Good Women* (1540), justifying both women's participation in civic life and their prowess in arms. The old tune was sung by the Scots reformer John Knox in his *First Blast of the Trumpet against the Monstrous Regiment of Women* (1558); for him rule by women, defects in nature, was a hideous contradiction in terms.

The confused sexuality of the imagery of female potency was not reserved for rulers. Any woman who excelled was likely to be called an Amazon, recalling the self-mutilated warrior women of antiquity who repudiated all men, gave up their sons, and raised only their daughters. She was often said to have "exceeded her sex" or to have possessed "masculine virtue"—as the very fact of conspicuous excellence conferred masculinity even on the female subject. The catalogs of notable women often showed those female heroes dressed in armor, armed to the teeth, like men. Amazonian heroines romp through the epics of the age—Ariosto's *Orlando Furioso* (1532) and Spenser's *Faerie Queene* (1590–1609). Excellence in a woman was perceived as a claim for power, and power was reserved for the masculine realm. A woman who possessed either one was masculinized and lost title to her own female identity.

THE PROBLEM OF SPEECH. Just as power had a sexual dimension when it was claimed by women, so did speech. A good woman spoke little. Excessive speech was an indication of unchastity. By speech, women seduced men. Eve had lured Adam into sin by her speech. Accused witches were commonly accused of having spoken abusively, or irrationally, or simply too much. As enlightened a figure as Francesco Barbaro insisted on silence in a woman, which he linked to her perfect unanimity with her husband's will and her unblemished virtue (her chastity). Another Italian humanist, Leonardo Bruni, in advising a noblewoman on her studies, barred her not from speech but from public speaking. That was reserved for men.

Related to the problem of speech was that of costume—another, if silent, form of self-expression. Assigned the task of pleasing men as their primary occupation, elite women often tended toward elaborate costume, hairdressing, and the use of cosmetics. Clergy and secular moralists alike condemned these practices. The appropriate function of costume and adornment was to announce the status of a woman's husband or father. Any further indulgence in adornment was akin to unchastity.

THE PROBLEM OF KNOWLEDGE. When the Italian noblewoman Isotta Nogarola had begun to attain a reputation as a humanist, she was accused of incest—a telling instance of the association of learning in women with unchastity. That chilling association inclined any woman who was educated to deny that she was or to make exaggerated claims of heroic chastity.

If educated women were pursued with suspicions of sexual misconduct, women seeking an education faced an even more daunting obstacle: the assumption that women were by nature incapable of learning, that reasoning was a particularly masculine ability. Just as they proclaimed their chastity, women and their defenders insisted on their capacity for learning. The major work by a male writer on female education—that by Juan Luis Vives, *On the Education of a Christian Woman* (1523)—granted female capacity for intellection but still argued that a woman's whole education was to be shaped around the requirement of chastity and a future within the household. Female writers of the following generations—Marie de Gournay in France, Anna Maria van Schurman in Holland, and Mary Astell in England—began to envision other possibilities.

The pioneers of female education were the Italian women humanists who managed to attain a literacy in Latin and a knowledge of classical and Christian literature equivalent to that of prominent men. Their works implicitly and explicitly raise questions about women's social roles, defining problems that beset women attempting to break out of the cultural limits that had bound them. Like Christine de Pizan, who achieved an advanced education through her father's tutoring and her own devices, their bold questioning makes clear the importance of training. Only when women were educated to the same standard as male leaders would they be able to raise that other voice and insist on their dignity as human beings morally, intellectually, and legally equal to men.

THE OTHER VOICE. The other voice, a voice of protest, was mostly female, but it was also male. It spoke in the vernaculars and in Latin, in treatises and dialogues, in plays and poetry, in letters and diaries, and in pamphlets. It battered at the wall of prejudice that encircled women and raised a banner announcing its claims. The female was equal (or even superior) to the male in essential nature—moral, spiritual, and intellectual. Women were capable of higher education, of holding positions of power and influence in the public realm, and of speaking and writing persuasively. The last bastion of masculine supremacy, centered on the notions of a woman's primary domestic responsibility and the requirement of female chastity, was not as yet assaulted—although visions of productive female communities as alternatives to the family indicated an awareness of the problem.

During the period 1300–1700, the other voice remained only a voice, and one only dimly heard. It did not result—yet—in an alteration of social patterns. Indeed, to this day they have not entirely been altered. Yet the call for justice issued as long as six centuries ago by those writing in the tradition of the other voice must be recognized as the source and origin of the mature feminist tradition and of the realignment of social institutions accomplished in the modern age.

We thank the volume editors in this series, who responded with many suggestions to an earlier draft of this introduction, making it a collaborative enterprise. Many of their suggestions and criticisms have resulted in revisions of this introduction, although we remain responsible for the final product.

PROJECTED TITLES IN THE SERIES

Isabella Andreini, *Mirtilla,* edited and translated by Laura Stortoni

Tullia d'Aragona, *Complete Poems and Letters,* edited and translated by Julia Hairston

Tullia d'Aragona, *The Wretch, Otherwise Known as Guerrino,* edited and translated by Julia Hairston and John McLucas

Francesco Barbaro et al., *On Marriage and the Family,* edited and translated by Margaret L. King

Francesco Buoninsegni and Arcangela Tarabotti, *Menippean Satire: "Against Feminine Extravagance" and "Antisatire,"* edited and translated by Elissa Weaver

Rosalba Carriera, *Letters, Diaries, and Art,* edited and translated by Catherine M. Sama

Madame du Chatelet, *Selected Works,* edited by Judith Zinsser

Vittoria Colonna, Chiara Matraini, and Lucrezia Marinella, *Marian Writings,* edited and translated by Susan Haskins

Princess Elizabeth of Bohemia, *Correspondence with Descartes,* edited and translated by Lisa Shapiro

Isabella d'Este, *Selected Letters,* edited and translated by Deanna Shemek

Fairy Tales by Seventeenth-Century French Women Writers, edited and translated by Lewis Seifert and Domna C. Stanton

Moderata Fonte, *Floridoro,* edited and translated by Valeria Finucci

Moderata Fonte and Lucrezia Marinella, *Religious Narratives,* edited and translated by Virginia Cox

Catharina Regina von Greiffenberg, *Meditations on the Life of Christ,* edited and translated by Lynne Tatlock

In Praise of Women: Italian Fifteenth-Century Defenses of Women, edited and translated by Daniel Bornstein

Lucrezia Marinella, *L'Enrico, or Byzantium Conquered,* edited and translated by Virginia Cox

Lucrezia Marinella, *Happy Arcadia,* edited and translated by Susan Haskins and Letizia Panizza

Chiara Matraini, *Selected Poetry and Prose,* edited and translated by Elaine MacLachlan

Alessandro Piccolomini, *Rethinking Marriage in Sixteenth-Century Italy,* edited and translated by Letizia Panizza

Christine de Pizan, *Debate over the "Romance of the Rose,"* edited and translated by Tom Conley and Virginie Greene

Christine de Pizan, *Life of Charles V,* edited and translated by Nadia Margolis

Christine de Pizan, *The Long Road of Learning,* edited and translated by Andrea Tarnowski

Madeleine and Catherine des Roches, *Selected Letters, Dialogues, and Poems,* edited and translated by Anne Larsen

Oliva Sabuco, *The New Philosophy: True Medicine,* edited and translated by Gianna Pomata

Margherita Sarrocchi, *La Scanderbeide,* edited and translated by Rinaldina Russell

Gabrielle Suchon, *"On Philosophy" and "On Morality,"* edited and translated by Domna Stanton with Rebecca Wilkin

Sara Copio Sullam, *Sara Copio Sullam: Jewish Poet and Intellectual in Early Seventeenth-Century Venice,* edited and translated by Don Harrán

Arcangela Tarabotti, *Convent Life as Inferno: A Report,* introduction and notes by Francesca Medioli, translated by Letizia Panizza

Laura Terracina, *Works,* edited and translated by Michael Sherberg

Katharina Schütz Zell, *Selected Writings,* edited and translated by Elsie McKee

"On gracious God relying, / My skillful hand applying, / Devoted deeds allying," motto of Justine Siegemund's *Court Midwife*.

VOLUME EDITOR'S INTRODUCTION

THE OTHER VOICE

Justine Siegemund[1] never had children. This circumstance left her with the time and energy to devote most of her adult life to the demanding profession of midwifery. It could also have barred her from the profession altogether, for in the seventeenth century in the German territories those who wished to serve as midwives were expected to have borne children themselves. Thus when in 1690 Siegemund, now an experienced, respected, and unusually well-connected practitioner, published her *Court Midwife*, a handbook of midwifery that was to go through seven subsequent editions[2] and thus make her by far the most famous German midwife of her era and a uniquely articulate teacher of and spokesperson for the skilled and reflective midwife, she felt compelled to argue her case once more. A physician, she explained, does not have to suffer every disease he treats to be a successful physician; why then should a midwife be unable to deliver babies merely because she herself has borne none?

Even as Siegemund justified her failure to meet seventeenth-century norms for practitioners of midwifery, her entry into print culture required

1. There are several variants of Siegemund's name, including Sigmund, Sigismund, and Siegmund, several of which will be seen in the translation itself. As was customary in seventeenth-century German, the feminine ending "in" is often added as well: Siegemundin. Justine is also variously spelled with an *e* or an *a*. In the handbook itself Siegemund spells the name with an *a*. "Justine Siegemund," the variant used in this introduction and in all the notes to the translation, is the standard spelling for cataloging the handbook internationally.

2. Jean M. Woods and Maria Fürstenwald, *Schriftstellerinnen, Künstlerinnen und gelehrte Frauen des deutschen Barock. Ein Lexikon*, Repertorien zur deutschen Literaturgeschichte, 10 (Stuttgart: J. B. Metzler, 1984), list an edition of 1692 as the second German edition (118), which I cannot confirm. They, however, overlook the edition of 1724 recorded by Waltraud Pulz, *"Nicht alles nach der Gelahrten Sinn geschrieben": Das Hebammenleitungsbuch von Justina Siegemund*, Münchner Beiträge zur Volkskunde, 15 (Munich: Münchner Vereinigung für Volkskunde, 1994), 177.

additional vindication. Indeed, when she wrote her handbook, there was quite simply no precedent for medical writing by women in Germany. At best there was the German translation of the handbook (1609, German 1619) by the French midwife Louise Bourgeois (1563–1636), a text that, as we shall see, differs in character from Siegemund's. Although birth was still largely managed by midwives in Germany, printed instruction in midwifery was the prerogative of male authors. Indeed, in the German territories and across Europe in general, medical men were turning out a growing number of handbooks of gynecology and obstetrics.

As Siegemund attempted to defend her book, her gender placed her in a double bind. The format and composition of her book pointedly, indeed aggressively, invoke the authority of her own personhood and experience, an experience that she had been able to acquire *because* she was a woman. Yet this same personhood potentially undercut her authority, for women writers occupied a severely circumscribed space not merely in medicine but in the print culture of early modern Germany in general. The few women who did publish in German in the seventeenth century tended to make apologies for their work, even to suggest, as did one of the premier German poets of the century, Catharina Regina von Greiffenberg, that they published only because male authorities told them to.[3]

Having chosen to take the unconventional step of putting her work as a midwife in the public eye, Siegemund vigorously set about seeking ways of protecting her book from anticipated criticism. She obtained and carefully displayed in the book official religious and medical approval. Moreover, she called on women whom she had served, both the powerful and the humble, for their patronage and testimony, electing ultimately to include testimonials from eleven of her female clients that had been generated years earlier by her legal difficulties with a Liegnitz town doctor. Furthermore, she apologizes more than once in her book for her simple, unlearned—and implicitly female—style of writing. At the same time, she uses precisely this perceived weakness to establish the authenticity of her work. "The manner of the book will sufficiently show that it is my work," she slyly insists, "for I am accustomed to speaking about things as I find them daily in my profession and I have recorded everything here as I speak of it."

Competition and conflict with other midwives, barber-surgeons, and university-trained doctors surface in her text, and the book itself fueled that competition and conflict. As we shall see, her book came under attack by a Leipzig surgeon and professor of anatomy shortly after its publication. Un-

3. Catharina Regina von Greiffenberg, *Sieges-Seule der Busse und Glaubens* [1675], vol. 2 of *Sämtliche Werke*, ed. Martin Bircher and Friedhelm Kemp (Millwood, NY: Kraus Reprint, 1983), sig.):(8^{v}.

daunted, Siegemund met the professor's accusations head on—that is, once she had them translated from the Latin she could not read.

Among pre-eighteenth-century European books on obstetrics, Siegemund's text constitutes a rarity not merely on account of its female authorship, but also because of its depiction of practical expertise. Through years of work as a midwife, through trial and error and study, Siegemund had become in the early modern context the next-best thing to an expert in obstructed labor and had perfected techniques of cephalic and podalic version.[4] She could thus write a text based and focused on her practical experience of birth, as opposed to the largely theoretical knowledge of most university-trained men.

The visibility of Siegemund as writer, obstetric practitioner, and proponent of a professionalism based on observation, reflection, and experience accorded her a status virtually unknown to German women in this period. She had learned to accept limitations imposed on midwives by the medical establishment, yet in her handbook she negotiates those limitations to show herself to considerable advantage. As a midwife, Siegemund relies on her hands and their sense of touch, her experience, and her ingenuity. Her *Court Midwife,* in describing thoughtful, reasoned, and innovative approaches to obstructed labor, belies the simple myth of a gender divide between the unscientific and traditional female practitioner on the one hand and the scientifically trained and progressive male professional on the other.

LIFE AND WORK[5]

Like many of the men who shaped culture in the German territories in the seventeenth and eighteenth centuries, Justine Siegemund, née Dittrich,[6] was

4. "Cephalic version" means manually turning the fetus in the uterus to make it possible to extract the fetus by the head; "podalic version" means manually turning the fetus in the uterus to make it possible to extract the fetus by the feet.

5. The following biographical sketch and discussion of Siegemund's *Court Midwife* draws on information that Siegemund herself supplies, documentation included in later editions of the *Court Midwife,* as well as my own work published in *Signs* in 1992 (Lynne Tatlock, "Speculum Feminarum: Gendered Perspectives on Obstetrics and Gynecology in Early Modern Germany," *Signs* 17 (summer 1992): 725–60); it is further greatly indebted to Waltraud Pulz's painstaking pursuit and documentation of the traces of Siegemund's life in her 1994 book. Pulz was not only able to substantiate some of Siegemund's own statements and to track down additional facts, but had access to a rare funeral oration by Daniel Bandeco, court chaplain at Brandenburg (Daniel Bandeco, *Die von Gott zu Gott gezogene Kinder Gottes / Bey Christlicher Beerdigung und Volckreicher Begleitung Frauen Justinä Sigmundin / Gewesenen Königl. Preußischen und Chur-Brandenburgischen Hoff-Wehmutter,* Berlin 1705). As Pulz points out, the exact dating of events in Siegemund's life is difficult, as Siegemund's account does not always line up with Bandeco's biographical sketch, and both accounts are occasionally contradicted by archival evidence.

6. I am using the spelling of Siegemund's maiden name used in the first edition of the handbook. The name has several alternate spellings.

born into a Lutheran pastor's family and doubtlessly owed her ability to read and write to this circumstance. In keeping with Protestant custom, she acquired the literacy necessary to read the Bible and run an orderly household.[7] Her birth on December 26, 1636, in Rohnstock (Polish: Roztoka) in what was then the ethnically German territory of Jauer (Polish: Jawor) in Silesia, now a part of Poland, brought her into a violent world that in Central Europe was witnessing the end of the second decade of the Thirty Years' War, a world shaped by religious strife.

Since 1618 the German territories had waged war against one another, with the Habsburg archduke, King of Hungary, and Holy Roman Emperor, united in the person of Ferdinand II (1619–37, Holy Roman Emperor 1619–37) and subsequently Ferdinand III (1608–57, Holy Roman Emperor 1637–57), leading the Catholic Counter-Reformation against Protestant rulers. Protestantism had made deep inroads into the German territories, even among the nobility in hereditary Habsburg lands in present-day Austria. It had had an especially deep impact in Silesia, in what would become heavily contested territory over the next century until in 1763 most of Silesia fell definitively to Prussia. In the seventeenth century, the Habsburgs not only strove to reestablish Catholicism in their hereditary lands but pushed into Silesia where, upon the death in 1675 of the last male heir of the ducal line, Georg Wilhelm, at age fifteen, Leignitz-Brieg-Wohlau fell to the imperial crown of Leopold I (1640–1705, Holy Roman Emperor, 1658–1705). Precisely in Silesia, this most sorely tried of German territories, German letters, Protestant and urban in origin, flourished as they had nowhere else in the German-speaking world for decades. The Silesian city of Breslau (Polish: Wroclaw), in particular, distinguished itself as the locus of what would come to be known as the Silesian School of Poets. Of course Siegemund, whose interests and vocation lay elsewhere, did not participate directly in this sophisticated literary life, yet her readiness to dispute in writing and to participate in print culture suggests origins in a for-the-time unusually literate culture.

As a Lutheran pastor, Siegemund's father, Elias Dittrich, found himself besieged during his short lifetime, and soon after Siegemund's birth he was on the run. After having suffered many trials and tribulations at the hands of imperial agents undertaking the re-Catholicizing of Silesia, he died in 1650, two years after the Peace of Westphalia officially ended the Thirty Years'

7. In 1643, a protestant contemporary, Hans Michael Moscherosch, advises precisely this kind of literacy for girls in his *Insomnis Cura parentum*.

War. Siegemund's mother, the daughter of a so-called free gardener,[8] was left with the task of rearing her thirteen-year-old daughter alone. On October 10, 1655, not yet nineteen years of age, Siegemund married Christian Siegemund, a steward and secretary, who over the course of his lifetime filled various such posts. Their marriage would last forty-two years, although Siegemund may have lived apart from him after his run-in with the law in 1673;[9] certainly, her successful practice of the profession of midwifery afforded her a high degree of mobility outside domestic confines.

Siegemund writes in her autobiography of her suffering at age twenty at the hands of ignorant midwives who mistakenly thought she was pregnant. This torturous experience, she claims, aroused her curiosity and led her to read books on obstetrics. Three years later, she responded to a desperate call from a local midwife, who knew of her reading, to intervene in a case of obstructed labor resulting from an arm presentation. Her success in this case, or rather, as she later admits, her dumb luck, led the midwife to request her assistance in many more cases. For twelve years thereafter she gathered experience among the village poor, never expecting payment for her services, she notes, and certainly not expecting to earn a living as a midwife. Over the course of these twelve years she established a reputation for her skill in difficult births and eventually served noblewomen, as well as women from the urban middle classes. Unlike many of her peers, Siegemund enjoyed the luxury of ample time, freed as she was from bearing and rearing children, and the luxury of working without remuneration, as her husband, who was gainfully employed, could support her.

Siegemund nowhere mentions a formal apprenticeship of the kind required by some German cities, although she does casually refer to these twelve years as her "apprenticeship." Her careful scripting in her handbook of the instruction and testing of a novice midwife by an experienced midwife attempts to regularize the training of midwives in a way that she herself had not experienced it. Nevertheless, she insists throughout her book on the importance of experience and reflection on it, that is, on a habit of mind—and not specific information—that must be acquired through praxis and that she herself acquired as an autodidact.

Around 1670 she was offered a post as a town midwife in Liegnitz

8. A "free gardener," possessing a small plot of land, falls somewhere between a cottager, who owns no land and works for others, and a land-owning peasant (Pulz, *Das Hebammenleitungsbuch von Justina Siegemund,* 27).

9. Pulz, *Das Hebammenleitungsbuch von Justina Siegemund,* 46.

(Polish: Legnica), a date she marks as the beginning of her career proper. In this position she was eventually consulted about the grave illness of Duchess Luise von Anhalt-Dessau, the regent and widow of Duke Christian von Liegnitz-Brieg-Wohlau, who had died on February 28, 1672. Siegemund describes in the introduction to her handbook how, after being brought in by the physicians attending the duchess in her illness, she successfully removed a tumor from the duchess's cervix. Presumably as a result of this successful operation, Siegemund served the court from 1672 to the duchess's death in April of 1680, although she also continued to attend the confinements of the townswomen of Liegnitz, women in neighboring villages, and nobility in the surrounding countryside.

Not surprisingly, professional success elicited enmity. The conflict of the midwife with her one-time supervisor, Martin Kerger, presents a dramatic case in point. Perhaps emboldened by the fact that Siegemund had lost her protector, Duchess Luise, Kerger, a Liegnitz town doctor, denounced Siegemund in court in 1680, accusing her of using harmful remedies and of employing dangerous methods of inducing labor and accelerating births to suit her convenience. She was in short, he alleged, a reckless and avaricious practitioner who manipulated the process of birth in order to serve as many women as possible. This rancorous case, which is partially documented in chapters 7 and 8 of Siegemund's manual, stretched out over four years and led Siegemund to seek retirement from her public office. Nevertheless, she defended herself aggressively. Eventually, the witnesses whom Kerger had thought to cite against her testified in her favor, and judgments sought from the medical faculties of Frankfurt on the Oder, Leipzig, and Jena confirmed her innocence.

The favorable judgments of the medical faculties turned on the fact that Kerger had accused her of acts that were in themselves impossible: (1) cutting veins in the uterus to hasten the birth—Siegemund correctly insisted that no such parturition-inducing veins exist—and (2) separating the placenta from the uterus at eight months—Siegemund pointed out the impossibility of reaching the placenta before dilatation of the cervical os has occurred. Moreover, Kerger's accusation that she ruptured the membranes of the amniotic sac to hasten birth did not stick either, inasmuch as Siegemund argued that she employed this procedure only to keep the fetus in a favorable position for birth, not to accelerate the birth. Perhaps because of this unpleasant experience, in her handbook she later repeatedly cautioned midwives against rupturing the membranes frivolously.[10] Furthermore, although

10. Rupturing the membranes of the amniotic sac is still employed in modern Western medicine to help induce labor. Siegemund's argument that she does not employ this procedure to ac-

midwives were increasingly discouraged from using medicines, which were ever more monopolized and regulated by male apothecaries and university-trained doctors, the medical faculty of Frankfurt on the Oder judged that Siegemund was not wrong to use the common remedies that she had employed to stop hemorrhaging and combat lassitude in newborn babies. Siegemund, who now was in a position to bring a countersuit, instead dropped the case. She was no longer able to pursue it long distance; indeed, her new employer, none other than the Elector of Brandenburg, had forbidden her from doing so.

In 1683, in the midst of her battle with Kerger, Siegemund had taken up a post in Berlin-Cölln at the court of Frederick William, Elector of Brandenburg. Here she was obliged to attend not only confinements at the court but also those of other women in need of her services in Berlin and its environs. In Brandenburg too, she found herself confronted with restrictions on her practice, surveillance, and competition and envy on the part of doctors as well as other midwives, for example, the midwife Elske Blanker, who had served the court since 1673 and who launched various complaints against Siegemund.[11] Presumably as a result of such professional trials and tribulations, in the handbook Siegemund models prudent management of the birthing room, in which her midwife avoids conflicts with other midwives where possible, judiciously consults male doctors, and defers to male rivals as a matter of expediency.

As an official court midwife in Brandenburg, Siegemund was sent to other courts as well, particularly those allied with the House of Hohenzollern. She, for example, attended Elector Frederick III's sister, Marie Amalie, Duchess of Saxony-Zietz, at the birth of her children, Frederick William (1690), Karoline Amalie (1693), Sophia Charlotte (1695), and Frederick August (1700). She was also lent to the court of August the Strong in Dresden to serve the Saxon Electress Eberhardine at the birth of her son Friedrich August II (1696).[12] In her handbook she alludes to her presence in January 1689 at the birth of Maria Amalia, the fourth child of Henriette Amalie, consort of Heinrich Casimir II von Nassau-Dietz, Statthalter of Friesland.

While in Holland in 1689, Siegemund recounts, she received encouragement from, among others, Mary II of England to publish her notes on midwifery. In fact, we can surmise that she was by then already well on her way toward publishing a carefully composed book. While in Holland, she

---

celerate labor rests on somewhat shaky grounds, although it is clear enough that she means to say that she ruptures the membranes only when an obstetric complication makes it necessary.

11. Pulz, *Das Hebammenleitungsbuch von Justina Siegemund,* 54.

12. Pulz, *Das Hebammenleitungsbuch von Justina Siegemund,* 55.

commissioned and paid for thirty-nine engravings that illustrated positions of the fetus in the uterus and the manipulations required to extract it and one engraving that illustrated three steps in the removal of a tumor, as well as foldout copies of two illustrations from contemporary books of anatomy (see below) and a foldout illustration of a combination birthing chair and bed and of the instruments that she recommended for the practice of midwifery. These forty-three illustrations, together with her portrait, which she commissioned from Samuel Blesendorf (1633–1706) in Berlin, promised an important, albeit expensive book.

Having once determined to publish, Siegemund set about protecting herself and her rights to her book. In the weeks before Easter in 1689 she journeyed from The Hague to Frankfurt on the Oder to submit her manuscript to the review of the medical faculty there. The faculty's favorable testimonial would be reproduced with each edition of her manual. She also saw to it in April 1689 that she received an attestation from the chaplains at the Brandenburg court to the religious orthodoxy of her work. Like the testimonial from the medical faculty, this document appears in all subsequent editions. In April, May, and June of the same year she procured printing privileges from the Elector of Brandenburg, the Holy Roman Emperor, and the Elector of Saxony, respectively. In the absence of copyright, such privileges served to guarantee her right to her work in the territories in question and promised to punish anyone who should dare to produce or sell pirated editions. The ten-year guarantees from the Holy Roman Emperor and the Elector of Saxony expired without consequence to her during her lifetime. The twenty-year privilege in Brandenburg became moot upon her death. In Berlin in 1708, three years after her death and eighteen years after the original publication, a Berlin printer brought out a second edition of the handbook.

Not surprisingly, the publication of Siegemund's *Court Midwife* roused the ire of some members of the male medical establishment. Despite the inclusion of the testimonial from the medical faculty of Frankfurt on the Oder or perhaps because of this attestation from a rival university, a Leipzig professor, Andreas Petermann (1649–1703), regarded the publication of the book as sufficient provocation to accuse Siegemund of presenting techniques of delivery that were speculative and absurd.

The public conflict began with a corollary in a Latin thesis on gonorrhea by the student Tobias Peucer, written under the direction of Petermann, that contained the gratuitous statement that much of what was in the book *The Court Midwife* was pure speculation and that it was incomprehensible how the book had passed the review of the medical faculty of Frankfurt on the

Oder. When these remarks came to her attention, the by-now well-seasoned midwife did not shy away from controversy. Siegemund had the accusations against her translated and thereafter complained to the university of Leipzig, learning only then that Petermann was the author of the corollary. At that point she elected to publish a response to it with Liebpert, the court printer who had printed her book.

Petermann, who now openly admitted that he had authored these remarks, published a further paper specifying his objections to the book. Even as he made the lofty argument that Siegemund presented manipulations that were dangerous and detrimental to the common weal, his frequent citing of his own case—and thus his prerogative as university professor, surgeon, and physician—suggests that he also believed Siegemund to have transgressed invisible gender boundaries by publishing this handbook in the first place. Indeed, in her presentation of the successful extraction of fetuses without the use of surgical instruments in cases where a male doctor or surgeon might have been called in to perform a craniotomy or embryotomy, Siegemund had most certainly trodden on contested territory.

Siegemund defended herself by returning to the stratagem she had employed in her book, that is, she reminded her readers that her intention in writing the handbook was not to address the learned scholarly community, but rather to instruct midwives. She, moreover, demanded to know specifically what Petermann considered absurd, stressing that the fact that manipulations were seldom performed did not mean they were ineffective. In medicine, she pointed out, many practices that had once been unusual and obscure were later praised, inasmuch as medicine was continually progressing. And, she emphasized once again, everything in her book had proven effective in her practice. Of course, she conceded, if proven in error, she would be happy to be instructed otherwise.

As Siegemund herself asserted in her defense, Petermann had not understood her book properly. He, for example, confused finger palpation of the cervix with the introduction of the entire hand into the uterus.[13] Petermann's insistence on his own expertise, furthermore, became somewhat ludicrous when he based it on his professional experience of delivering close to ninety infants, whereas at the time Siegemund, as she herself emphasized, had delivered over five thousand.

13. See, e.g., Siegemund, *Die Königl.-preußische und chur-brandenburgische Hof-Wehemutter* (1741; reprint, Hildesheim: Georg Olms, 1976), 279. Pulz stresses that in doing so he confuses two precise terms on which Siegemund had based her entire manual (*Das Hebammenleitungsbuch von Justina Siegemund*, 113).

In the end the conflict between Petermann and Siegemund, as documented in later editions of the handbook, shifted to one between Petermann and the University of Frankfurt on the Oder. Petermann's blunt questioning of the probity of the medical faculty of Frankfurt on the Oder in putting its seal of approval on Siegemund's book in fact raises the interesting question of whether the medical faculty, which did not systematically teach obstetrics, had sufficient knowledge to vet the handbook in the first place.[14] At the same time, precisely a sense of his own (male) university prerogative led Petermann to an aggressive judgment of a text he himself had not understood properly; by claiming for himself practical knowledge of birth of which he had relatively little in comparison to the average midwife, he was, after all, carving out a new sphere of activity for male university-trained physicians and encroaching on traditionally female territory.

Siegemund appears in any case not to have suffered from, but instead to have profited by this public attack. In the 1690s she received additional privileges from the Brandenburg court including permission to have a sedan chair made in which she could be transported as needed.[15]

By the time of her death on November 10, 1705, a month and a half before her sixty-ninth birthday, Siegemund had acquired stature and wealth exceptional for a midwife in the German territories in this period. Among other things, she owned a house in Berlin.[16] Daniel Bandeco, the deacon and minister of Berlin's second oldest parish church, the *Marienkirche*, commemorated her with a funeral oration in which he noted that she had left her wealth to good causes in Brandenburg and Saxony. Siegemund had, he remarked in the oration—which as a further indicator of status also appeared in print—delivered 6,199 infants over the course of her career, among these twenty princely births.[17]

Had Siegemund not written *The Court Midwife*, this remarkable career would have been quickly forgotten; the handbook, however, gained for her a modest place in German medicine and German medical history. She came to be associated in particular with the so-called double-handed podalic ver-

14. Pulz, *Das Hebammenleitungsbuch von Justina Siegemund*, 111.

15. Pulz, *Das Hebammenleitungsbuch von Justina Siegemund*, 58. Pulz points out that the sedan chair was a recent innovation in Brandenburg, brought about by French Huguenots who had settled there after the revocation of the Edict of Nantes in 1685.

16. Pulz notes that Siegemund owned a house in Berlin as of 1690 (*Das Hebammenleitungsbuch von Justina Siegemund*, 46).

17. Pulz, *Das Hebammenleitungsbuch von Justina Siegemund*, 59. See note 5 above.

sion, which facilitated the extraction of the infant, dead or alive.[18] Moreover, for a time in Germany a technique of cephalic version of the fetus in the womb bore her name: the Siegemund-Busch version.[19]

"THE COURT MIDWIFE" (1690)

Justine Siegemund took advantage of her connections to several reigning houses of seventeenth-century Europe when she elected to publish her handbook of midwifery. She had it printed at her own expense by the Brandenburg court printer, a printer located in the elector's castle in Cölln.[20]

The title page names her affiliation with the Brandenburg court, an affiliation reaffirmed in some copies of the first edition of the book by a dedication to Electress Sophie Charlotte. Other copies of the text supply alternate dedications, calling variously on other noble and royal women of the time: Mary II, Queen of England; Anna Sophia, Electress of Saxony; Henriette Amalie, Princess of Nassau; and Charlotte, Duchess of Schleswig-Holstein,[21] and Siegemund alludes pointedly on the title page and in her introduction to the "highborn persons" who persuaded her to publish her jottings on the art of midwifery. In the five alternate dedications Siegemund mentions her gratitude to these powerful women for their support of her work, thereby seeking their further patronage. She also quite openly expresses her hope that their noble names will pave the way for greater receptiveness in various territories to the book, in other words, for greater sales.

Yet for any contemporary familiar with Louise Bourgeois's handbook of midwifery, the title page of Siegemund's handbook will have raised false expectations with its suggestion of electoral affiliation and access. While Siegemund did try to exploit the cachet of patronage, her book tells nothing about her work at the Brandenburg court or indeed much about her ministering to exalted personages. When she alludes to serving Duchess Luise

18. Beatriz Spitzer, *Der zweite Rosengarten. Eine Geschichte der Geburt* (Hannover: Elwin Staude, 1999), 223. Hans Schadewaldt probably refers to this type of version when he writes of "a certain handhold" ("Die Frühgeschichte der Frauenheilkunde," *Zur Geschichte der Gynäkologie und Geburtshilfe: aus Anlaß des 100 jährigen Bestehen der Deutschen Gesellschaft für Gynäkologie und Geburtshilfe,* ed. L. Beck, 89–94 [Berlin: Springer, 1986], 91–92).

19. F. von Winckel, ed. *Handbuch der Geburtshülfe,* 3 vols. (Wiesbaden: J. F. Bergmann, 1903–7), 3 (pt.1): 210.

20. The location of the oldest settlement in modern-day Berlin and the area now considered the center of Berlin.

21. She was also known as Charlotte (Karolina / Carolina) of Silesia.

in Silesia, she refers to her anonymously as a "highborn person." In her discretion vis-à-vis her noble clients, she contrasts sharply with Bourgeois, who provided a vivid account of the birth of Louis XIII of France and his brothers and sisters.

While Siegemund alludes to her later access to reigning houses, she explains that her practice began when she was called to the bedside of a peasant woman. The patients she specifically identifies according to economic and social class include mainly peasant women and women who, roughly speaking, belonged to the urban middle class. A number of women who testified for her in court in Liegnitz are identified by name with their husband's trade, e.g., miller, tawer, farmer, cooper, cabinetmaker.

As mentioned above, *The Court Midwife* evidences an attempt on the part of its author to protect her publication from attacks from all sides. The testimonial from three chaplains from the Brandenburg court anticipates objections on religious grounds. The critical role that midwives played in delivering babies and the high mortality rates of both parturient mothers and infants had long put the activity of midwives in close proximity to religious matters, and Siegemund's work as midwife would have constituted no exception. Obstetric handbooks by men insist that the midwives be pious and warn against deviation from this rule. Midwives who published of course needed to be mindful of this expectation. Bourgeois's piety is, for example, elegantly expressed in her handbook by the cross she wears in the engraved portrait bound near the front of her book.

Siegemund too is at pains to demonstrate her piety. In addition to furnishing the volume with the court chaplains' certification, she repeatedly stresses that her professional success and indeed the outcome of any birth depend on God's help. The book, furthermore, opens with a Bible verse pertaining to pious midwives and ends with an expression of humility before God: "Honor is God's alone." In the various dedications Siegemund asks for God's blessing for her patrons or expresses her intention to pray on their behalf. The text addresses readers as Christians, and the fictitious interlocutors are identified as Christian as well. The very name of the novice midwife—Christina—makes obvious the religious affiliation. Nevertheless, while, like Bourgeois, Siegemund shows that she is firmly rooted in Christianity, she avoids belaboring the negative connotations of child bearing in the Christian context or rehearsing the redemptive value of the same.[22]

Many early modern gynecological and obstetric treatises and hand-

22. In "The Estate of Marriage," for example, Martin Luther had preached that women would be redeemed in childbirth but that they might pay for this redemption with their lives.

books commence by reminding readers of Genesis 3:16 in which God condemns all women to bear children in pain as a result of Eve's disobedience.[23] While Siegemund also quotes this verse, she does not do so until chapter 3, and then she passes over it quickly, reminding her readers simply that while everyone suffers pain, there is no explanation as to why one woman experiences more difficulties than another. When in chapter 1, Christina asks whether all women have a womb, Justina answers simply that all women must have a womb if they are to bear children; she does not suggest that women's sole purpose is to bear children nor does she waste words preaching to women about redemption through childbirth.[24] She does warn her readers throughout, however, that childbirth will not go well without God's help.

The Court Midwife stands out among early modern European handbooks by midwives for its attempt at systematic presentation of detailed information on the management of difficult births. Siegemund presents in this book the most common complications of labor, that is, malpresentations of many kinds including presentation of the umbilical cord and placenta previa and their solutions with specific examples from her practice. The tidy division of the book into nine chapters according to subject matter, or complications, and the inclusion of an index underline the intention of systematic pedagogy and instruction.

The book does not reproduce the medical lore and humoral medicine that, for example, constitute a large portion of *The Midwives Book or the Whole Art of Midwifry Discovered* (1671) by Jane Sharp and that led Sharp to begin a book on "the art of midwifery" with a description of the penis. Unlike Sharp's book and Bourgeois's famous tome, Siegemund's manual does not provide information on fertility, management of pregnancy, or care of the newborn, nor does it present a random collection of cases like Sarah Stone's *A Complete Practice of Midwifery* of 1737, which exhibits little attempt at arrangement by subject matter, let alone systematic instruction. Nor is it, like the long unpublished notebook of the Frisian midwife Catharina Schrader (1693–1745), a chronicle of her cases.

In *The Court Midwife* Siegemund creates a dialogue between two midwives in which information is presented clearly and reasonably systemati-

23. See, e.g., Cornelis Solingen, *Embryulcia oder Herausziehung einer Todten-Frucht durch die Hand des Chirurgi*, 2d ed. (Wittenberg, 1712), 3.

24. Note, for example, that the first German version of Bourgeois quotes 1 Tm 2:15: "Yet woman will be saved through bearing children, if she continues in faith and love and holiness, with modesty" (Louise Bourgeois, *Ein gantz new / nützlich und nohtwendig Hebammen Buch* (Oppenheim, 1619), 20).

cally in question and answer form for the purpose of teaching midwives and potentially also their clients and (male) supervisors. In part 1, a novice midwife, Christina, poses questions that are answered by an experienced midwife, aptly named Justina (the author's representation of herself). In part 2, in order to reiterate the information imparted in part 1, Justina catechizes Christina in a turnabout that Siegemund justifies as sound pedagogy.

In her handbook Siegemund successfully adapts an established form of high cultural writing to her own purposes. In early modern Europe the dialogue, the idiom of both urbane philosophy and the New Science, was the format used by learned men. James Wolveridge, for example, had employed the dialogue in *Speculum Matricis* (1671), but in that dialogue, a male doctor instructs a female midwife.

The dialogue in both parts of Siegemund's book imitates conversation, with the two midwives responding to one another with exclamations, chiding, expressions of gratitude, requests for more information, admissions of failure to understand, and apologies for returning to subjects previously discussed. While a nineteenth-century German medical historian criticizes Siegemund's style as chatty,[25] precisely this style, indeed its quasi-literariness, differentiates the book from similarly structured obstetrical texts from the era. In 1715 the Swedish doctor Johan von Hoorn, for example, offered a dialogue between the midwives, Pua and Siphra, names derived from midwives named in the Old Testament, but then failed ever to use the names anywhere but the title page. The questions and answers are simply labeled as such and show no attempt to imitate conversation. If Hoorn had taken his cue from Siegemund's successful book, he had missed the point. Siegemund's naturalistic dialogue offers not merely information but models of reflection on obstetric practice that invite imitation.

Siegemund's use of dialogue also recalls the official catechism or the test that many cities used to license midwives, in which the exchange occurs between the aspiring midwife and the city authority, and such catechisms may well have suggested the dialogue form to Siegemund in the first place. A later handbook of midwifery from 1735, marketed under the name of the experienced Augsburg midwife Barbara Widenmann but actually written by her husband, offers such a catechism to consumers as a means of verifying the skills of the midwife whom they wish to engage. Like Hoorn's manual, Widenmann's handbook does not attempt to personalize the questions and answers as a conversation and misses the opportunity to model the seeking of and reflection on advice.

25. E[duard] C[aspar] J[akob] Siebold, *Versuch einer Geschichte der Geburtshilfe*, 2d ed. (Tübingen: Franz Pietzschen, 1901–4), 2:202.

The forty-three illustrations, to which Siegemund repeatedly refers in her explanations, form an integral part of the book and support its pedagogical intent. The engravings depicting the postures and manipulations of the fetus are, however, not particularly well executed; in their depiction of the capaciousness of the uterus, they somewhat contradict the text, where Siegemund reminds her readers that the fetus is always curled up and where she compares the child in the womb to a person in a wet chemise or a piece of meat in a sack and hence very difficult to turn.

Two of the engravings that Siegemund commissioned in Holland, however, display remarkable artistry. Siegemund attributes one of them, an engraving of the uterus and ovaries, to Reinier de Graaf (1641–73) but does not identify the source of the second one. This second one, a depiction of the fetus curled up in the womb with the layers of the abdomen and the uterus peeled back, is copied from an engraving in *Anatomia humani corporis* (1685) by the Dutch anatomist, Govard Bidloo (1649–1713).[26] Siegemund may have had the opportunity to view this book when she traveled to The Hague in 1689, and Bidloo himself could have numbered among the Dutch doctors whom she mentions having met there, inasmuch as he had been appointed to the professorship of anatomy at The Hague in 1685.

Like Bourgeois, Siegemund used her own portrait as a frontispiece to her book. Bourgeois's and Siegemund's prefacing of their obstetric handbooks with an image of themselves demands a closer look, for it raises central issues inherent in the ascent of women to fame within print culture and the subsequent transmission of their writing.

The first German translation of Bourgeois's handbook contains an apology in verse for the shortcomings of the portrait, an apology that expands the sentiment of the four lines in French under the portrait in the original French edition. Bourgeois's portrait, the apology asserts, can only depict the outward features. It cannot depict the subject's virtue, art, skill, and intelligence. These features are to be found only in the writing itself, as this writing speaks where the silent portrait cannot. The apology thus proposes that the portrait invites us to read the words reproduced in the book; it creates the wish to discover what lies behind the face. And Bourgeois's face is a self-conscious and intelligent one in this German rendition. Her body is erect, her eyes look slightly to her left, and a slight amusement plays around her month, as if she had stories to tell. At forty-five, she still looks relatively young.

Siegemund, by contrast, has the slightly relaxed and tired face of a much older woman, her portrait having been engraved at the time of the

26. The chief artist for Bidloo's book was Gerard de Lairesse (1640–1711).

book's appearance when she was about sixty-three years of age. She appears somewhat stooped; her eyes peer off to her right, lending her face a thoughtful look. She wears a dignified scarf and robe that appear to be of rich fabric—lace, braid, perhaps velvet. The oval frame of the portrait rests on a pedestal that identifies the subject by name and occupation.[27]

The portrait in and of itself functions as a sign of authority and success, an authority that foregrounds and draws on personhood. The character and prestige of the historical person—as constituted by the very existence and inclusion of the portrait—thus initially justify and guarantee the contents of the text. Subsequently, that text and the image conjoin to represent and affirm retroactively the historical person, placing the articulating subject squarely in the center. The focus on the writer herself is especially marked in both of these highly self-representational texts where the midwife is the eyewitness and chief actor and where her practical experience looms large.

Both Bourgeois's and Siegemund's portraits would, however, eventually become detached from their texts as medical science superseded the information their books offered, and their portraits would thus become silent icons to be reproduced in the rare medical history. Nevertheless, during the decades in which these texts still circulated widely, these authors, as visible writing subjects, enjoyed a public stature extraordinary for women.

Authority in Siegemund's text rests at least as much with women as with men. In the text proper Justina—Siegemund's alter ego—her female witnesses, and her female pupil, Christina, constitute the loudest authoritative voices. When Siegemund calls on male authority, she does so largely to frame her book and to justify and protect her practice. Her thinly disguised account of a historical conflict between a doctor and a midwife presents one such instance.

When in chapter 8 Christina asks Justina to tell her about the remedies she has found useful in childbirth, the question provokes a vehement reaction. Justina reminds her that she knows well enough what happened to the midwife Titia whom a certain doctor Sempronius accused of inappropriately using remedies. Justina then proceeds to outline the entire conflict, one that actually belonged to Siegemund's own history with Kerger—the contemporary reader could probably easily surmise Titia's true identity even if Sempronius's identity remained a mystery. To corroborate Justina's account of the unjustified suit, Siegemund turns to the institutionalized medicine of men, supplying affidavits from the medical faculties of Frankfurt on the

27. Later editions replace this information with the rhyme that appeared on a separate page opposite the portrait in the first edition.

Oder, Leipzig, and Jena, that is, copies of those from her own case with the names changed. The inclusion of these documents serves once again to anticipate criticism and substantiate Siegemund's authority; at the same time, the case induces midwives to limit their practice to manual intervention. Medicines—even everyday remedies—belong to the purview of men.

Siegemund notes elsewhere in the text the dangers presented by the use of powerful medicines during birth. Indeed, Maria Thym testifies to the disastrous effects of a powder given to her by none other than Dr. Kerger, whom Justina had consulted. The powder, Thym claims, led to such violent labor pains that she nearly lost her reason. Kerger had not, however, bothered to stay around to observe or manage the effects. Similarly, Barbara Vogt testifies that the midwife, Frau Corneliuß, gave her medicine that caused her to labor so hard she broke her birthing chair. As it turns out, the fetus lay in the womb with its neck broken, pushing into the birth canal, presumably as a result of this inappropriate stimulation of the labor pains.

This and other testimonies included in chapter 7 constitute one of the remarkable features of the manual. Siegemund introduces these testimonies *in her own name* in the chapter before the one in which her alter ego, Justina, tells the story of Titia and Sempronius. Within the chronology of the text she therefore has no need to maintain a fiction that has not yet been introduced.

The testimonies are, for the most part, signed and dated. Some of them include place names and the names of other midwives, and two of them openly name Kerger. The text drops even the notion of the alter ego, Justina, by preceding the testimonies with the sworn statement of court officials in Liegnitz who declare that "on this date there appeared before us at the ordinary court the honorable and virtuous Frau Justina Siegemund, née Dittrich, the famous and experienced midwife of these parts." The oral testimonies of nine of the women are paraphrased in the third person by the court officials. The two noblewomen, who, unlike the other witnesses, are not identified by name, speak in their own words in the first person, having provided written attestations rather than oral testimony.

While scholars of early modern Germany have pointed out the treacherous and collusive role female accusers and witnesses played in the trials of women denounced as witches, the testimonies included in Siegemund's manual show women's part in a legal proceeding in a more favorable light, that is, women reveal themselves to be sober and discriminating consumers of obstetric services who cannot be induced to testify against a midwife whose work they admire and whose services they desire. They do not blame Siegemund when their babies die, but rather insist that she did everything possible under the circumstances, indeed, that she saved their lives. They stress, furthermore, that they owe the lives of their few surviving children to her skill. They make clear that Siegemund consulted them and their relatives and attendants, as well as doctors and other midwives. Regina Titz, for example, reports that after she had experienced several stillbirths Siegemund even suggested that she might want to try working with someone else. The testimonial of an unidentified noblewoman states quite simply that everyone wants to use Siegemund—if only they could get her.

The accounts provided here describe labor that lasts not hours but days, of outcomes that are favorable because the mothers—not the infants—survive without permanent bodily injury. The birthing room is a noisy place where the patient, her attendants and relatives, the midwife, the doctor, and other midwives express opinions and sometimes disagree. Siegemund talks to her patients, explains to them what she is going to do, and shows them what has gone wrong. The parturient mother is by no means silent or silenced and may even end up mediating a quarrel between the midwives. Problems of transportation and communication dictate that the midwife often arrives after the fetus has died and the mother has labored for days. The parturient mothers exhibit patience in their acceptance of a degree of trial and error in their deliveries and the inevitability of pain and loss and repeated pregnancies.

It may surprise the modern reader to learn that Siegemund regards delivering the mother of her child, alive or dead, as a good outcome. In fact the very necessity of separating mother and child stands at the center of her handbook. What is more, delivering the mother does not mean necessarily that she will survive, but for the mother to have any chance at all of survival, the midwife must extract the infant.

Siegemund sometimes employs instruments, namely, the crotchet or hook, to remove the fetus. She also occasionally amputates a limb in order to facilitate extraction. Yet she stresses that in extracting children largely through manual manipulation she differs from male rivals who regularly perform embryotomies, that is, dissection of the fetus within the womb. She differs, furthermore, from these male practitioners in eschewing the use of the speculum, which she believes causes women unnecessary pain. Years of practice enable her to know by touch, and thus she does not need to see. She assures aspiring midwives that they too will come to know the female body in the same way. They must learn to think about what their hands tell them.

The illustrations that Siegemund provides do not, as one might be tempted to conclude, essentially contradict this position. Rather, the engravings enable Siegemund's pupils to interpret what they encounter, when undertaking the critical finger palpation or when inserting their hand fully into the uterus, and to make sound judgments accordingly. Siegemund does not include them with the idea that anyone will actually be able to see a fetus in the womb as though through an invisible wall, but to help midwives and their clients understand the challenges of manual manipulation.[28] In stressing in chapter 5 the importance of touch, as opposed to the gaze through the speculum, Siegemund resists the reliance on sight that marked seventeenth-century university medicine, necessitated of course by men's lack of practical experience in touching living female bodies.[29]

Although even today obstetricians and midwives may palpate the abdomen externally to determine the position of the fetus, Siegemund makes no mention of regularly undertaking external examinations and even warns that they can deceive the midwife. She insists rather that the finger palpation of the cervix constitutes the one truly reliable source of information. In foregrounding the "touch," as it was then called, she was insisting on the midwife's time-honored prerogative to touch her patient's genitals, a prerogative that doctors and man-midwives would eventually erode. Over the course of the manual, however, it becomes clear that women do not necessarily readily submit to this procedure, even when it is undertaken by a midwife, and Siegemund becomes somewhat polemical on the subject, blaming the women for their foolishness, her complaint oddly paralleling that of male doctors who have even less access to living women's interior.

Although Siegemund displays a sense of competition with some male practitioners, she shows interest in the approval of the medical establishment as represented by university-trained physicians. She describes how in the presence of doctors she is able to perform a craniotomy with her tiny hand to the chagrin of a French barber-surgeon who has also been called to the scene and who is first unable to use the instrument she lends him and is then disadvantaged by his large hands. Her techniques, she explains, have evolved from experience, study, reflection, and judgment, and she can point to steady progress during the thirty-odd years of her practice in accruing knowledge.

Siegemund continually exhorts midwives to reflect on what they are doing. Reflection is key here to the development of a professional ethic. She thus allies herself with reason and in this sense participates in public and

28. In the seventh testimony in chapter 7 Siegemund describes her demonstration to a parturient woman and her attendant of how the stillborn child had lain in the womb to explain the difficult birth.

29. Male university-trained physicians were familiar with female anatomy through dissection and the proliferating anatomical handbooks, that is, through dead bodies or drawings of them.

male-dominated discourses of her era.[30] Yet women had long been thought and would continue to be thought to lack precisely the capacity to reason.

THE AFTERLIFE OF "THE COURT MIDWIFE"

In 1691 an unsigned Dutch translation, entitled *Spiegel der Vroed-Vrouwen* (The midwives' mirror), appeared in Amsterdam accompanied by an eighty-four-page obstetric manual by the Dutch physician and surgeon Cornelis Solingen (1641–87).[31] The inclusion of Solingen's instruction in obstetrics initiates a trend in the publication history of Siegemund's book by which male doctors and surgeons publicly and in print, subtly and not so subtly, begin to circumscribe Siegemund's accomplishment.

Solingen's *Korte Instructie wegens het Ampt ende Pflicht der Vroed-Vrouwen* (Short instruction concerning the office and duty of midwives) had first been published in Amsterdam in 1684. Although Solingen displays here a patronizing attitude toward midwives in general and takes ignorant midwives to task for pretending to expertise, he also shows himself open to working amicably with them and even expresses admiration for those whom he considers the better among them. In some respects his manual and Siegemund's complement and support one another. On the other hand, points of agreement notwithstanding, the physical coupling of Siegemund's book with Solingen's work in and of itself undermines a central feature of the former, that is, the education of women by a woman and the foregrounding of women as experts in a field of activity that is exclusively theirs.

The explanation by the Dutch bookseller of his decision to publish these two texts back-to-back supports the impression that the inclusion of Solingen's book diminishes Siegemund's. He reports that when he gave the translation to certain Dutch physicians for review, they advised him to publish Solingen's manual along with it. While the bookseller certainly had no

30. Ulrike Gleixner points out the increasing emphasis by clients on skill and the qualifications of the midwife over the course of the eighteenth century, whereas in the previous two centuries clients had spoken solely of God's mercy and grace. She sees this major shift in the cultural history of midwifery as largely propelled by the medical establishment ("Die 'Gute' und die 'Böse': Hebammen als Amtsfrauen auf dem Land (Altmark/Brandenburg, 18. Jahrhundert)," In *Weiber, Menscher, Frauenzimmer: Frauen in der ländlichen Gesellschaft, 1500–1600,* ed. Heide Wunder and Christina Vanja, 96–122 [Göttingen: Vandenhoeck and Ruprecht, 1996], 114). Siegemund in fact demonstrates an investment in precisely the expertise shared by male practitioners but then deviates somewhat from her male contemporaries in her understanding of the particulars of this expertise.

31. We must assume that the translation was unauthorized. Certainly it appeared outside the jurisdiction of the three printing privileges that Siegemund had obtained for her book.

reason to wish to discredit or contradict a book he wished to sell, the addition of Solingen's text inadvertently does so. We cannot but suspect that the Dutch physicians whom he consulted felt the need to have the authoritative voice of one of their own to balance that of a German woman.

And indeed, an odd turn of events suggests how favorably Solingen's text stacked up against Siegemund's in the eyes of at least some members of the German medical establishment. Solingen's manual was translated into German in 1693 by none other than Tobias Peucer, the author of the Latin thesis to which Petermann had added his scornful corollary. Exhibiting an unabashed rivalry with and antipathy toward midwives, Peucer opens his translation of Solingen with the complaint that women do not allow male physicians access to them and that midwives encourage them in this noxious reticence, resulting in the parturient mothers' death or infertility.[32]

A comparison of Siegemund's book with Solingen's brings into sharp relief the contested territory of midwifery. Siegemund's responsible midwife takes care not to overreach her authority and consults a doctor as necessary, just as Solingen advises. Siegemund, furthermore, emphatically warns midwives against using medicines, in keeping with Solingen's admonition, but in doing so she stresses the troubles that the use of medicines can bring midwives rather than the doctors' superior knowledge of them per se. Solingen spells out midwives' obligation to fetch a doctor to perform a caesarean section in cases when the mother has expired and the fetus is believed to be still viable. Siegemund, on the other hand, does not mention this surgical procedure. She alludes merely to her dilemma as to how to proceed when death for either the fetus or the mother is imminent if drastic action is not taken. Given her expertise and her dependence on the good will of her female clients, we can infer that the question for her can in the end only be whether or not to save the mother at the expense of the child. Solingen, who in 1673 had published his *Embryulcia,* a treatise on the extraction of dead fetuses by the *hand of the surgeon,* had, however, clear ideas about the limits of the midwife's activity that conflicted with those of Siegemund, whose fifth chapter is devoted precisely to the removal of dead fetuses.

Siegemund herself offers tart remarks on the subject of incompetent midwives, but she is not interested in their character as such; she dwells rather on the knowledge, habits of learning, and experience that she believes are necessary for them to acquire. Although Solingen too advocates the education of midwives, he, by contrast and in keeping with the practice

32. Tobias Peucer, introduction to *Hand-Griffe der Wund-Artzney nebst dem Ampt und Pflicht der Weh-Mütter,* 2d ed. (Wittemberg, 1712), sig.)(2-)(3.

of his male contemporaries, specifies the necessary character and circumstances of the midwife, a topic he takes up on the first page of his manual. She should be neither too old nor too young, gentle, kind, loyal, discreet, jolly, friendly, intelligent, reasonable, careful, and cunning. She should have sympathy for the suffering of others. She should not be given to drink or to gluttony, nor should she object to getting out of bed in the middle of the night. Midwives should, furthermore, neither curse nor gossip and should know how to encourage their patients when the time comes. By naming the characteristics of the good midwife Solingen simultaneously evokes early modern stereotypes of the bad midwife—lazy, greedy, drunken, gossipy, flighty, and irresponsible—the strange woman who at times of birth had access to houses she otherwise would never have entered. Moreover, Solingen's recommendation in his introductory remarks that midwives be married, have their own children, but not be pregnant themselves quite simply undercuts Siegemund's introductory self-justification.

Citing none other than Louise Bourgeois in this prefatory address to midwives, Solingen bluntly evokes the atrocities of midwives who harbor unwed mothers, women impregnated by married men or non-Christians, and cites as well the case of a midwife who infected her patients and their families with syphilis. Siegemund, by contrast, maintains a discreet silence on the larger sociology of midwifery and the abuses associated with it and focuses instead on the technical aspects she wishes to impart to her fellow midwives.

Solingen's manual not only dictates the behaviors and attributes of midwives, but also seeks to exercise control over the mothers themselves. Siegemund too alludes to the difficulties recalcitrant patients present to the midwife and her ability to manage births and emphasizes the need for the patient to let herself be governed by the competent midwife. Nevertheless, she offers no specific instruction about how parturient mothers should comport themselves except to remind them that they need to thank God for their delivery. Solingen, by contrast, prescribes correct behaviors for mothers during pregnancy, parturition, and the lying-in period. Confinement is here an apt word, for women are supposed to limit their activity: not eat too much, not travel or ride horseback, avoid coughing, refrain from wearing high heels and tight lacing, guard against certain smells and loud noises, spare themselves emotional excitement of any kind, and keep quiet. During delivery itself they are to be silent, gentle, obedient, and patient.

In some respects, Siegemund's text exhibits greater expertise and more critical thinking. For example, in the case of the treatment of placenta previa, the only recourse Solingen can suggest is to remove the placenta as quickly as possible, while Siegemund, by contrast, correctly recommends piercing the placenta to stay the bleeding. Furthermore, while Siegemund discredits the superstitious practice of untying all knots in the parturient mother's clothing, Solingen perpetuates the superstition by advising the untying of knots without offering any rationale for doing so.

After Siegemund's death, virtually unaltered versions of the manual appeared in 1708 in Berlin and in 1715 and 1724 in Leipzig. Beginning with a new edition published in Berlin in 1723, all other German editions circumscribe the original text in ways that distort its original import and impact. In some respects the supplements in these editions enhance the authority and usefulness of the text, but, as with the coupling of the Dutch translation with Solingen's manual, often at the cost of discrediting or contradicting the message of Siegemund's original book and certainly at the cost of minimizing her achievement as a woman.

In 1701 Frederick III of Brandenburg had crowned himself King in Prussia. Johann Daniel Gohl (1675–1731), the editor of the edition of 1723, thus felt justified in altering the original title of Siegemund's book to read *The Royal Prussian and Electoral Court Midwife* instead of *The Electoral Court Midwife*. Nevertheless, even as Siegemund's title became more lofty, her original text was literally surrounded by the additions, opinions, warnings, and preaching of male editors and doctors.

Gohl strikingly altered the appearance of the text by adding marginalia designating the topics under discussion to guide the reader through the book, marginalia characteristic of learned texts by men of the day. By seeing to it that the book conformed ever more to such external standards, he both increased its authority and initiated its cooptation. Whereas Siegemund's original manual explicitly cites only one published male gynecological authority—de Graaf—Gohl inserts two references to François Mauriceau[33] and promotes himself from the title page on through the book as a superior authority. Gohl's new framing of the handbook underwent three more editions (1741, 1752, 1756), the editions of 1741 and 1752 serving as the basis for reprints in 1976 and 1980, respectively. Siegemund's manual was thus preserved for the present in its altered, learned, indeed masculinized, habit.

The absence of the customary characterological catalog of the good midwife had apparently caught Gohl's attention. He added to his *Royal Prussian Court Midwife* a second introduction invoking his authority as Assessor of the College of Medicine in Berlin. Here he enumerates the char-

33. Pulz, *Das Hebammenleitungsbuch von Justina Siegemund,* 106.

acterological traits of a "careful and reasonable midwife" and articulates a strong prohibition against abortion. In advocating literacy, at least minimal training in techniques of version and knowledge of anatomy, however, he speaks in the spirit of the original handbook.

Ignoring Siegemund's express advice that midwives avoid doling out remedies, Gohl also added as an appendix a previously published anonymous treatise on useful medicines, thirty-eight pages in length and directed in particular at those who lived in the countryside and thus did not have ready access to urban medicine.[34] Gohl himself was in fact the author of this text.[35] By adding information on medicine to Siegemund's handbook, he doubtlessly gained a larger audience for the publication and also increased its currency, even as he undercut one of its central points.

Not content with these alterations and additions, however, Gohl also reproduced the controversy generated by the handbook by including five public documents (eighty-four additional pages) resulting from the confrontation between Petermann and Siegemund. This documentation includes an eight-page refutation by Siegemund (dated January 29, 1692), Petermann's eighteen-page paper against Siegemund's book (dated November 13, 1691), an undated thirty-page refutation by Siegemund, a brief including a set of documents from the University at Frankfurt on the Oder justifying the testimonial featured in Siegemund's book and refuting Petermann (dated October 17, 1691–March 15, 1692), and finally Petermann's reaction to the brief from the University of Frankfurt on the Oder. Inasmuch as the documents are included without further commentary, it is not entirely clear that Siegemund comes off as the unquestioned victor in this dispute. Indeed, the sequence of the documents literally gives Petermann the last word.

The manual won Siegemund emulators in addition to detractors. Anna Elisabeth Horenburg's thin handbook, published in Wolfenbüttel in 1700, shows in its organization and presentation signs of its author having read Siegemund. Horenburg, who had had eight children by the time she thought of becoming a midwife, inserts Anna, her alter ego, into the text as an experienced midwife. In two parts, one on so-called natural births and one on difficult and "unnatural" births, Anna answers Margaretha's questions in a conversational manner that recalls Siegemund's handbook. Horenburg's text,

34. *Einige Fürsichtige Lehren und Heilsame Artzney-Mittel / Denen auf dem Lande wohnenden und insgemein unerfahrnen Kinder-Müttern zum Besten* [1720]. Pulz points out that midwives who practiced in the countryside needed to be able to provide their patients more than the services Siegemund describes and, in particular, with medicines (*Das Hebammenleitungsbuch von Justina Siegemund,* 106).

35. Pulz, *Das Hebammenleitungsbuch von Justina Siegemund,* 106.

however, deviates immediately from Siegemund's by cataloging up front the characteristics of a good midwife. Siegemund's handbook may also have inspired the aforementioned Widenmanns to publish their handbook in 1735.

The engravings of the positions of the fetus in the womb included in Siegemund's handbook, lacking in artistry though they may be, experienced a surprising afterlife in the eighteenth century. Lorenz Heister's *Chirurgie* (Surgery, Nuremberg 1731); a posthumous edition of Guillaume Mauquest de La Motte's (1655–1737) *Traité complet des accouchemens naturels, non naturels, et contre nature* (Thorough treatise on natural and unnatural births, 1721), which appeared in 1765; Johannes Fatio's *Helvetisch-vernünftige Wehe-Mutter* (Helvetianly sensible midwife, written 1690, first published 1752); and Johann Friedrich Schütze's *Gründliche Anweisung zur Hebammenkunst* (Thorough instruction in the art of midwifery, 1770) all show clear evidence of having borrowed from these copper engravings. In the posthumous edition of La Motte's book, the manipulations and handholds are, however, presented as the accomplishment of surgeons, not of midwives.[36]

PRINCIPLES OF TRANSLATION

Present-day native speakers of German—even those with medical training—would not find *The Court Midwife* an easy read. The German language has changed substantially in the over three hundred years that have elapsed since the book's publication and, moreover, in 1690 neither the language nor spelling had been standardized. As contemporaries themselves noted, German prose remained clumsy and had not yet developed the power and precision of expression of the vernacular languages of neighboring rival nations. *The Court Midwife* certainly reflects this state of affairs. In keeping with the style of writing of the time, Siegemund's sentences are very long, and the logical relation between the subordinate clauses is often nebulous by modern standards. Punctuation, spelling, and case endings deviate from modern standard German, and the text displays consistency in its inconsistency with regard to all three features. The lexicon differs somewhat from modern standard German, as do the meanings of some still commonly used words. Moreover, the book was printed in Fraktur, that is, Gothic typescript, a typeface that in its various guises dominated German printing through

36. Christine Loytved, *Dem Hebammenwissen auf der Spur. Zur Geschichte der Geburtshilfe. Katalog zur Ausstellung der Arbeitsgruppe Gesundheitswissenschaften (AGW)—Gesundheits- und Krankheitslehre, Psychosomatik,* Schriften der Universitätsbibliothek Osnabrück, 6 (Osnabrück: Universität Osnabrück, Arbeitsgruppe Gesundheitswissenschaften, 1997), 20–21.

1945. Modern-day German readers are no longer practiced in reading Fraktur and need time to work their way into the text.

Perhaps most bewildering of all, Siegemund's conception of the body and her understanding of the process of birth differ from contemporary Western conceptions and understanding, and it is therefore not always obvious what she is describing. Like her seventeenth-century English contemporaries, she may, for example, refer to the womb and mean merely the uterus or she may mean both the uterus and the vagina. Moreover, when she speaks of birth, she does not distinguish the stages of birth identified by modern medicine nor does she describe descent as stages of passage through the pelvis; indeed, she claims that she is unable to know anything about the pelvis because it is covered with flesh. Nevertheless, her frequent mention of the pubic bone makes clear that she is performing manipulations that facilitate precisely such passage.

Yet even if *The Court Midwife* presents difficulties for the present-day German reader, Siegemund's contemporaries would have found the practical, nonacademic language of the book reasonably easy to follow. No doubt they did not find the syntax of the handbook quite as uncontrolled and murky as does the modern reader. With this original accessibility in mind, I strove first and foremost for readability. To this end, I broke up many of the longer sentences, smoothed out the language here and there, and made the syntax more logical than it is in the original. Nevertheless, I also attempted to preserve a sense of the foreignness of the place, epoch, and culture of this text to our present-day North American English-speaking medicalized world. I therefore tried to avoid obviously modern expressions and the medicalized technical language, common today in the vocabulary of even nonspecialists, employing instead standard seventeenth-century English expressions for parts of the body and for aspects of birth. The language of this translation is thus much indebted to the handbooks of Jane Sharp and Sarah Stone, as well as to other seventeenth- and eighteenth-century English obstetrical texts with whose language I familiarized myself. I did not, however, attempt to reproduce seventeenth-century English per se; rather I strove to generate a mildly quaint and colloquial English that could evoke the origins of Siegemund's text in a distant time, place, and mind-set. Moreover, I sought to preserve the distinctness of this midwife's voice, the intelligent, assertive, unvarnished, and practical voice with which Siegemund speaks of her work as she finds it "daily in her profession."

VOLUME EDITOR'S BIBLIOGRAPHY

PRIMARY SOURCES

Bidloo, Govard. *Anatomia humani corporis.* Amsterdam, 1685.

Bourgeois, Louise. *Observations diverses sur la sterilité.* Paris, 1609.

———. *Ein gantz new / nützlich und nohtwendig Hebammen Buch.* Oppenheim, 1619.

Chapman, Edmund. *A Treatise on the Improvement of Midwifery.* 3d ed. London, 1759.

Churfürstliche Brandenburgische Medicinal-Ordnung und Taxa. Berlin, 1694.

Counsell, George. *The London New Art of Midwifry; or, The Midwife's Sure Guide.* London, 1758

Culpeper, Nicholas, and Alice Culpeper. *A Directory for Midwives; or, A Guide for Women in Their Conception, Bearing, and Suckling Their Children.* London, 1755. First published in 1651.

Deventer, Hendrik van. *Neues Hebammen-Licht.* 6th ed. Jena, 1775. First published as *Operationes chirurgicae novum lumen exhibentes obstetricantibus.* Lugduni Batavorum, 1701–33.

———. *New Improvements in the Art of Midwifery.* Translated by Robert Samber. London, 1728.

Ettner, Johann Christoph. *Des Getreuen Eckarths Unversichtige Heb-amme.* Leipzig, 1715.

Graaf, Reinier de. *De mulierum organis generationi inservientibus.* 1672. Reprint, Niewkoop: de Graaf, 1965.

Greiffenberg, Catharina Regina von. *Sieges-Seule der Busse und Glaubens.* Vol. 2 of *Sämtliche Werke.* Edited by Martin Bircher and Friedhelm Kemp. Millwood, NY: Kraus. Reprint, 1983. First published in 1675.

Grosses vollständiges Universal-Lexicon aller Wissenschafften und Künste. 64 vols. Halle and Leipzig: Johann Heinrich Zedler, 1732–54.

Habel, P. Antonio. *Der Weyland Durchlauchtigsten Fürstin und Frauen / Frauen Carolina, Hertzogin zu Schleßwig Holstein.* Breslau, 1708.

Hoorn, Johan von. *Die zwo um ihrer Gottesfurcht und Treue willen von Gott wohl belohnte Wehe-Mütter Siphra und Pua.* 2d ed. Stockholm and Leipzig, 1737.

Horenburg, Anna Elisabeth. *Wohlmeynender und Nöthiger Unterricht der Heeb-Ammen.* Hannover and Wolfenbüttel, 1700.

Moscherosch, Hans-Michael. *Insomnis Cura parentum: Christliches Vermächtnuß oder, Schuldige Vorsorg eines trewen Vatters.* Strassburg, 1643.

Luther, Martin. "The Estate of Marriage." In vol. 45 of *Luther's Works*, edited by Walther I. Brandt, 17–49. Philadelphia: Muhlenberg, 1962.

Mauriceau, François. *The Diseases of Women with Child, and in Child-Bed.* Translated by Hugh Chamberlen. 2d ed. London, 1683. First published as *Des maladies des femmes grosses et accouchées*. Paris, 1668.

Schrader, Catharina. *"Mother and Child Were Saved." The Memoirs (1693–1740) of the Frisian Midwife Catharina Schrader.* Translated and annotated by Hilary Marland. Nieuwe Nederlandse Bijdragen tot de Geschiedenis der Genesskunde en der Natuurwetenschappen. Amsterdam: Rodopi, 1987.

Sharp, Jane. *The Midwives Book, or the Whole Art of Midwifry Discovered.* Edited by Elaine Hobby. Women Writers in English 1350–1850. New York: Oxford University Press, 1999.

Siegemund, Justine. *Die Chur-Brandenburgische Hoff-Wehe-Mutter.* Cölln, 1690.

———. *Die Chur-Brandenburgische Hoff-Wehe-Mutter.* Leipzig, 1715.

———. *Die Königl.-preußische und chur-Brandenburgische Hoff-Wehe-Mutter.* Berlin, 1723.

———. *Die Königl.-preußische und chur-Brandenburgische Hoff-Wehe-Mutter.* Berlin, 1741. Reprint, Hildesheim: Georg Olms, 1976.

———. *Die Königl.-preußische und chur-Brandenburgische Hoff-Wehe-Mutter.* Berlin, 1756.

———. *Spiegel der Vroed-Vrouwen.* Amsterdam, 1691. First published as *Die Chur-Brandenburgische Hoff-Wehe-Mutter.* Cölln, 1690.

Solingen, Cornelis. *Embryulcia oder Herausziehung einer Todten-Frucht durch die Hand des Chirurgi.* 2d ed. Wittenberg, 1712. Originally published as *Embryulicia ofte afhalinge eenes dooden vruchts door de handt van den heel-meester* (s'Gravenhage, 1673).

———. *Hand-Griffe der Wund-Artzney nebst dem Ampt und Pflicht der Weh-Mütter.* Translated and with an introduction by Tobias Peucer. 2d ed. Wittemberg, 1712. Originally published as *Manuale operatien der chirurgie, beneffens het ampt en pligt der vroedvrouwen* (Amsterdam, 1684).

———. *Korte instructie wegens het Ampt ende Pflicht der Vroed-Vrouwen,* published in one volume with Justine Siegemund's *Spiegel der Vroed-Vrouwen.* Amsterdam, 1691.

Stone, Sarah. *A Complete Practice of Midwifery.* London, 1737.

Völter, Christoph. *Neueröffnete Hebammen-Schul/ Oder Nutzliche Unterweisung Christlicher Heb-Ammen und Wehe-Müttern.* 2d ed. Stuttgart, 1687.

Widenmann, Barbara. *Kurtze/ Jedoch hinlängliche und gründliche Anweisung Christlicher Hebammen.* Augsburg, 1735.

SECONDARY SOURCES

Aikin, Judith P. "Gendered Theologies of Childbirth in Early Modern Germany and the Devotional Handbook for Pregnant Women by Aemilie Juliane, Countess of Schwarzburg-Rudolstadt (1683)." *Journal of Women's History* 15, no. 2 (2003): 40–67.

———. "The Welfare of Pregnant and Birthing Women as a Concern for Male and Female Rulers: A Case Study." *Sixteenth Century Journal* 35, no. 1 (2004): 9–41.

Donnison, J. *Midwives and Medical Men.* New York: Schocken Books, 1977.

Duden, Barbara. *Geschichte unter der Haut: Ein Eisenacher Arzt und seine Patientinnen um 1730.* Stuttgart: Klett-Cotta, 1987. (English-language edition: Barbara Duden, *The Woman beneath the Skin: A Doctor's Patients in Eighteenth-Century Germany.* Cambridge, MA: Harvard University Press, 1991.)

Eccles, Audrey. *Obstetrics and Gynaecology in Tudor and Stuart England.* Kent, OH: Kent State University Press, 1982

Evenden, Doreen. *The Midwives of Seventeenth-Century London.* Cambridge: Cambridge University Press, 2000.

Fasbender, F. *Geschichte der Geburtshilfe.* Jena: Gustav Fischer, 1906.

Flügge, Sibylla. *Hebammen und heilkundige Frauen: Recht und Rechtswirklichkeit im 15. und 16. Jahrhundert.* Frankfurt am Main: Stroemfeld, 1998.

Gabbe, Steven G., Jennifer R. Niebyl, and Joe Leigh Simpson, eds. *Obstetrics. Normal and Problem Pregnancies.* 4th ed. New York: Churchill Livingstone, 2002.

Gleixner, Ulrike. "Die 'Gute' und die 'Böse': Hebammen als Amtsfrauen auf dem Land (Altmark/Brandenburg, 18. Jahrhundert)." In *Weiber, Menscher, Frauenzimmer: Frauen in der ländlichen Gesellschaft, 1500–1600,* edited by Heide Wunder and Christina Vanja, 96–122. Göttingen: Vandenhoeck and Ruprecht, 1996.

Grooss, K. S. *Cornelis Solingen: A Seventeenth-Century Surgeon and His Instruments.* Leiden: Museum Boerhaave, 1990.

Hacker, Neville F., and J. George Moore, eds. *Essentials of Obstetrics and Gynecology.* Philadelphia: W. B. Saunders Company, 1986.

Jocelyn, H. D., and Brian P. Setchell, trans. *Regnier de Graaf on the Human Reproductive Orders: An Annotated Translation of* Tractatus de virorum organis generationi inservientibus *(1688) and* De mulierum organis generationi inservientibus tractatus novus *(1672).* Oxford: Blackwell Scientific, 1972.

Labouvie, Eva. *Andere Umstände: Eine Kulturgeschichte der Geburt.* Cologne: Böhlau, 1998.

Loytved, Christine. *Dem Hebammenwissen auf der Spur: Zur Geschichte der Geburtshilfe: Katalog zur Ausstellung der Arbeitsgruppe Gesundheitswissenschaften (AGW)—Gesundheits- und Krankheitslehre, Psychosomatik.* Schriften der Universitätsbibliothek Osnabrück, 6. Osnabrück: Universität Osnabrück, Arbeitsgruppe Gesundheitswissenschaften, 1997.

Metz-Becker, Marita, ed. *Hebammenkunst gestern und heute: Zur Kultur des Gebärens durch drei Jahrhunderte.* Marburg: Johans Verlag, 1999.

Myles, Margaret F. *Textbook for Midwives with Modern Concepts of Obstetric and Neonatal Care.* 8th ed. Edinburgh: Churchill Livingstone, 1975.

Pulz, Waltraud. *"Nicht alles nach der Gelahrten Sinn geschrieben": Das Hebammenleitungsbuch von Justina Siegemund.* Münchner Beiträge zur Volkskunde, 15. Munich: Münchner Vereinigung für Volkskunde, 1994.

Ricci, James V. *The Development of Gynaecological Surgery and Instruments.* San Francisco: Norman Publishing, 1990.

Roper, Lyndal. *Oedipus and the Devil. Witchcraft, Sexuality and Religion in Early Modern Europe.* New York: Routledge, 1994.

Schadewaldt, Hans. "Die Frühgeschichte der Frauenheilkunde." In *Zur Geschichte der Gynäkologie und Geburtshilfe: Aus Anlaß des 100 jährigen Bestehen der Deutschen Gesellschaft für Gynäkologie und Geburtshilfe,* edited by L. Beck, 89–94. Berlin: Springer, 1986.

Siebold v., E[duard]. C[aspar]. J[akob]. *Versuch einer Geschichte der Geburtshilfe.* 2d ed. Tübingen: Franz Pietzschen, 1901–4.

Spitzer, Beatriz. *Der zweite Rosengarten: Eine Geschichte der Geburt.* Hannover: Elwin Staude, 1999.

Tatlock, Lynne. "Speculum Feminarum: Gendered Perspectives on Obstetrics and Gynecology in Early Modern Germany." *Signs* 17 (summer 1992): 725–60.

Towler, Jean, and Joan Bramall. *Midwives in History and Society.* Dover, NH: Croom Helm, 1986.

Thadden, Rudolf von. *Die brandenburgisch-preussischen Hofprediger im 17. und 18. Jahrhundert: Ein Beitrag zur Geschichte der absolutistischen Staatsgesellschaft in Brandenburg-Preussen.* Arbeiten zur Kirchengeschichte 32. Berlin: Walter de Gruyter, 1959.

Ulrich, Laurel Thatcher. *A Midwife's Tale. The Life of Martha Ballard, Based in Her Diary, 1785–1812.* New York: Albert A. Knopf, 1990.

Wiesner, Merry E. "Early Modern Midwifery: A Case Study." In *Women and Work in Preindustrial Europe,* edited by Barbara A. Hanawalt, 94–113. Bloomington: Indiana University Press, 1986.

———. "The Midwives of South Germany and the Public / Private Dichotomy." In *The Art of Midwifery: Early Modern Midwives in Europe,* edited by Hilary Marland, 77–94. London: Routledge, 1993.

Wille, F. C. *Über Stand und Ausbildung der Hebammen im 17. und 18: Jahrhundert in Chur-Brandenburg.* Abhandlungen zur Geschichte der Medizin und der Naturwissenschaften, 4. Berlin: Emil Bering, 1934.

Wilson, Adrian. *The Making of Man-Midwifery: Childbirth in England, 1660–1770.* Cambridge, MA: Harvard University Press, 1995.

Winckel, F. von, ed. *Handbuch der Geburtshülfe.* 3 vols. Wiesbaden: J. F. Bergmann, 1903–7.

Woods, Jean M., and Maria Fürstenwald. *Schriftstellerinnen, Künstlerinnen und gelehrte Frauen des deutschen Barock: Ein Lexikon.* Repertorien zur deutschen Literaturgeschichte, 10. Stuttgart: J. B. Metzler, 1984.

On gracious God relying,
My skillful hand applying,
Devoted deeds allying

THE COURT MIDWIFE OF THE ELECTORATE OF BRANDENBURG, THAT IS, A HIGHLY NECESSARY MANUAL ON DIFFICULT AND UNNATURAL BIRTHS, PRESENTED IN A CONVERSATION, NAMELY, HOW WITH DIVINE HELP A WELL-INFORMED AND PRACTICED MIDWIFE PREVENTS SUCH THINGS WITH INTELLIGENCE AND A SKILLED HAND OR, WHEN NECESSARY, CAN TURN THE CHILD BASED ON MANY YEARS OF PRACTICAL EXPERIENCE AND FOUND TO BE TRUE. NOW, HOWEVER, PUBLISHED, AT HER OWN EXPENSE, ALONG WITH AN INTRODUCTION, COPPER ENGRAVINGS, AND A REQUISITE INDEX, TO HONOR GOD AND TO SERVE HER NEIGHBOR AND AT THE MOST GRACIOUS AND FERVENT DESIRE OF MANY ILLUSTRIOUS HIGHBORN PERSONS. BY JUSTINE SIEGEMUND, NÉE DITTRICH, OF ROHNSTOCK IN SILESIA IN THE PRINCIPALITY OF JAUER WITH SPECIAL PRIVILEGES FROM THE HOLY ROMAN EMPIRE AS WELL AS THE ELECTORATES OF SAXONY AND BRANDENBURG.

Cölln on the Spree
Printed by Ulrich Liebpert, Court Printer of the Electorate of Brandenburg, 1690.

So God dealt well with the midwives.
And because the midwives feared God he built them houses.
—Ex 1: 20–21[1]

That is, He blessed them in their profession and rewarded their loyalty.

1. I have adhered to Siegemund's original German in translating this quotation from scripture as "he built them houses," which deviates from Luther's rendering of this passage, namely, "he blessed their houses." The English Revised Standard Version reads, "he gave them families." Siegemund's text omits the rest of the verse "and the people multiplied and grew very strong."

My most gracious Electress and Lady.[2]
To the Most High Princess and Lady,
Lady Sophie Charlotte,[3]

Marquise and Electress of Brandenburg, in Prussia, of Magdeburg, Jülich, Cleves, Berg, Stettin, Pomerania, of the Kashubians and Wends, also in Silesia, of Crossen and Schwiebus, Duchess, born Duchess of Brunswick and Lüneburg, Baroness of Nuremberg, Princess of Halberstadt, Minden and Cammin, Countess of Hohenzollern, the Mark and Ravensberg, Lady of Ravenstein and the lands of Lauenburg and Bütow, etc., etc., etc.

Most High Electress,
Most Gracious Electress and Lady,

Because the present book was written up under the protection and favor of the Electoral Sovereign Authority, I consider myself obliged to submit it first to your Electoral Highness and thus most humbly take the liberty of placing your great name in front of it as powerful protection with the request that your Electoral Highness most graciously deign to accept it and with the wish that your Electoral Highness may for a long time yet be a blessing upon the Electoral House and the entire land and may become a mother many times over. Your most humble servant wishes you this, your Electoral Highness, my most gracious Electress and Lady, from the bottom of her heart.

Justina Siegmund [*sic*]

2. I have yet to find a copy that contains all five of the dedications that follow, but I have nevertheless chosen to include all of them here so as to provide an overview of the noblewomen to whom Siegemund attempted to ally herself in print. The variants that I have found to be most common are dedicated either to Sophie Charlotte only or Anna Sophia only. The locations of the rare variants with the additional dedicatees cited below are indicated in the notes.

3. Sophie Charlotte (1668–1705) of Brunswick-Lüneburg, married Frederick of Brandenburg in 1684, became Electress of Brandenburg in 1688, and Queen in Prussia in 1701. Sophie Charlotte was known for her learnedness and talent as a musician and maintained a friendship with, among others, the philosopher Gottfried Wilhelm Leibniz (1646–1716). Siegemund was of course attached to the Brandenburg court and here appropriately seeks Sophie Charlotte as her patron.

To the most Illustrious Electress and Lady, Lady Anna Sophia[4]

Née Princess of Denmark, Norway, of the Wends and Goths, Electress and Duchess of Saxony, Jülich, Cleves and Berg, Schleswig, Holstein, Stormarn and Dithmarshes, Countess in Thuringia, Marquise of Meissen, also of the Upper and Lower Lausitz, Countess of Oldenburg and Delmenhorst. Etc., etc., etc.

To my most gracious Electress and Lady.
Most Illustrious Electress,
Most Gracious Lady

Because I have often had the privilege of serving my neighbor in your lands and at the seats of your courts and have experienced many a serious complication there and was often given cause to collect such things in a book and inasmuch as this book is now also favored with the privilege of the Electorate of Saxony, I believed myself most humbly obliged, before I made this book known in your lands, to present your Electoral Highness the first issue under your exalted Electoral Name that it may find all the more protection and be of still greater use. May your Electoral Highness graciously deign to accept it. May God continue to bless your Electoral Highness in body and in soul, hear your prayers and allow you to see the prosperity of your house and your children's children, peace unto Israel, for which thing your most humble servant, the undersigned Justina Siegmund [*sic*], prays daily to God on behalf of your Electoral Highness, my most gracious Electoral Highness and Lady.

4. Anna Sophia (1646–1717) of Denmark, daughter of Frederick III of Denmark, married Johann Georg III, Elector of Saxony (1647–91) in 1666; mother of August the Strong (1670–1733); known for her learnedness in spiritual and worldly affairs, her knowledge of foreign languages, and her piety. She maintained a friendship with the pietist Philipp Jacob Spener (1635–1705) and founded the protestant orphanage in Dresden. In 1696 Siegemund was lent to the court of August the Strong in Dresden to serve the Saxon Electress Eberhardine at the birth of her son Friedrich August II (Pulz, *Das Hebammenleitungsbuch von Justina Siegemund,* 55).

To the Most High Princess and Lady, Lady Mary[5]

Queen of Great Britain, France and Ireland, etc., etc.

To my Most Gracious Queen,
Most High Princess,
Most Gracious Queen,

I would not be so bold as to appear before Your Royal Majesty with this book, if your most gracious command to have this book printed were not still on my mind. So because Your Royal Majesty with your superior understanding judged back then that the work would serve the common weal, I wanted to bring this work to the common weal for its greater benefit under the noble name of Your Royal Highness. It is my most humble plea that Your Royal Majesty graciously deign to accept it,[6] along with the book, with the addition of a heartfelt wish that God bless Your Royal Majesty and bless you with children in that great kingdom to the glory of His name and the good of the Church. With this wish I remain Your Royal Majesty's

Most Humble Servant
Justina Siegmund [*sic*]

5. Mary II (1662–94), Queen of England, was the eldest daughter of James II and Anne Hyde. After the forced abdication of her father, she and her husband, William of Orange (1650–1702), were crowned corulers of England in April 1689, she as Mary II and he as William III. She died childless of smallpox five years later. Mary had lived many years in Holland, her husband's home, before returning to England. Siegemund apparently met her just months before she was crowned. One of the exemplars of the first edition of Siegemund's handbook held at the University of Chicago contains this dedication, along with a dedication to (Henriette) Amalia (see below).

6. That is, the dedication.

To the Most High Princess and Lady, Lady Amalia,[7]

Princess of Nassau Born Princess of Anhalt, Duchess of Saxony, Engern and Westfalia, Countess of Ascania, Katzenellenbogen, Vianden, Dietz and Spiegelberg, Lady of Beilstein, Zerbst and Bernburg etc. etc.

To my most gracious Princess and Lady,
Most High Princess,
Gracious Princess and Lady,

Your Highness is already familiar with this book, insofar as it recently found most gracious favor in your eyes, both when I was at The Hague and when I was in Lewarden, and you, along with Her Majesty, the reigning Queen of England, thought it useful to have it published and also recommended its publication through your venerable physicians-in-ordinary. Thus on account of my most humble obligation, I have made so bold as to attach Your Highness's name, along with that of Her Majesty the Queen of England, so as to enable this book entry into the lands where I am not known. May Your Highness thus deign graciously to accept this [dedication] and be assured that I will never fail most humbly to acknowledge the favor I have enjoyed from Your Highness and to pray for your noble prosperity. I am the most humble servant, of Your Highness, my most gracious Princess and Lady,

Justina Siegmund [*sic*]
Berlin, February 11, 1690

7. Henriette Amalie of Anhalt-Dessau (1666–1726); in 1683 she married her cousin Henry Casimir II of Nassau-Dietz (1657–96). Siegemund speaks in her introductory chapter of having served her "a year ago," thus alluding to her assistance in the birth of Henriette Amalie's second daughter, Maria Amalia (1689–1771), on January 29, 1689. This dedication appears, along with the dedication to Mary II, in one of the exemplars of the first edition of the handbook, held by the University of Chicago.

To the Most High Princess and Lady, Lady Charlotte,[8]

Espoused Duchess of Schleswig, Holstein, Stormarn, and the Dithmarshes, born Duchess in Silesia, of Liegnitz, Brieg, and Wohlau, Countess of Oldenburg and Delmenhorst.

To my most gracious Princess and Lady.
Most High Duchess,
Most Gracious Princess and Lady,

The favor of having lived a long time under the gracious protection and care of not only Your Highness's venerable mother of blessed memory, but of also most humbly serving Your Highness and of thus being able to get farther in my profession, obliges me to everlasting gratitude, to which I would like to testify publicly by presenting this book under Your Highness's noble name, trusting that Your Highness will most graciously deign to permit it so this book will have all the more credibility with my countrymen. Because God so miraculously led me to this art and I served my fatherland[9] first in it, I could think of no more proper way of introducing this book as a contribution to it than under your noble name. It is my most humble request that you most graciously accept this and my everlasting wish that God's grace constantly rule over you.

The humble servant of your Highness, my most gracious Princess and Lady,

Justina Siegmund [*sic*]
February 11, 1690

By the grace of God, we, Johann Georg III,[10] Duke of Saxony, Jülich, Cleves, and Berg, the Grand Marshall of the Holy Roman Emperor and Elector,

8. Charlotte, aka Karolina (1652–1707) of Liegnitz-Brieg-Wohlau, daughter of Luise of Anhalt-Dessau and Duke Christian of Liegnitz-Brieg-Wohlau, sister of Duke Georg Wilhelm (d. November 21, 1675), the last of the line of the Piasts of Liegnitz-Brieg-Wohlau and wife of Friedrich von Schleswig-Holstein-Sonderburg-Wiesenburg, from whom she was divorced in 1680. Siegemund was engaged at court in Silesia for nearly a decade by Charlotte's mother in the 1670s. The exemplar of the first edition of Siegemund's handbook held by the Bio-Medical Library of the University of Minnesota contains this dedication, along with the above dedication to Sophie Charlotte.

9. That is, Silesia, and more specifically the Duchy of Brieg-Liegnitz-Wohlau.

10. Johann Georg III (1647–91), Elector of Saxony (1680–91) and husband of Anna Sophia (see above, note 4).

Count in Thuringia, Marquis of Meissen, also of Upper and Lower Lausitz, Baron of Magdeburg, Princely Count of Henneberg, Count of the Mark, Ravensberg, and Barby, Lord of Ravenstein, etc., hereby make known that Justina Siegemund of Berlin most humbly gave us to understand that she intended to publish her treatise, entitled "The Midwife of the Electorate of Brandenburg" and furnished with copper engravings. Because she feared damage from pirated editions, she most obediently requested a privilege from us, which petition we now grant. Thus we intend that in the electoral territory of Saxony and the lands and dioceses incorporated therein no bookseller or printer shall within ten years of the date given below reprint this book, nor shall anyone, in case such reprints occur in other places, sell or market them on pain of loss of all reprinted copies and a fine of one hundred rhenish gold guilders for each pirated edition, half of which shall then go to our treasury, the other half of which shall fall to Siegemund. She, for her part, is obliged assiduously to correct the oft-mentioned book, to have it printed up most elegantly and to use good white paper for it as often as it is reprinted. She is obliged, at her own expense, to give to our Supreme Consistory eighteen copies of each printing and edition in advance of sale. Further, she shall not cede our privilege to anyone without our foreknowledge and permission. Thus we herewith command and order each and every one of our prelates, counts, lords, those from the knighthood and nobility, our responsible local government officials and magistrates, administrators, tax collectors, customs officials, city councilors, judges, wardens, village mayors, and all of our subjects and aliens enjoying a citizen's rights to protect and support said Siegemund with respect to our privilege for the ten years granted her as it concerns us. And should anyone violate it and should she petition the execution of it, we command the aforementioned to act upon it without delay, to collect the fine that has been set without delay, and to deliver up the pirated copies at the appropriate place. This constitutes our firm intention. We have documented and sealed this privilege with our electoral privy seal and signed it with our own hand.

So done and granted in Dresden, June 17, 1689.

Johann Georg Elector
(L. S.)[11]
Hans Ernst Koch
Theodorus Werner/*S.*[12]

11. *Locus signilli* (Latin): location of the seal.

12. *Secretarius* (Latin): privy counselor.

We, Leopold,[13] elected Holy Roman Emperor, by the grace of God at all times the Augmenter of the Empire, in Germany, Hungary, Bohemia, Dalmatia, Croatia, and Slavonia, etc., King, Archduke of Austria, Duke of Burgundy, Styria, Carinthia, Carniola, and Würtemberg, and Count of Tyrolia, do with this letter publicly confess and make known to each and everyone that Justina Siegemund, sworn midwife[14] to the court of the Electorate of Brandenburg, gave us to understand that she had written down her many observations and experiences of thirty years in difficult births and the turning of children in unnatural postures. Further, she made known that with the encouragement of a number of princely physicians-in-ordinary and doctors as well as other persons she had resolved to publish them at her own expense as experiences that had been found highly useful, along with requisite copper engravings that would be useful to women giving birth, under the title "The Court Midwife of the Electorate of Brandenburg, or a Highly Necessary Manual on Difficult and Unnatural Births." Inasmuch as this work necessitated great expenditure and it is to be feared that pirated editions would make it difficult in all kinds of ways for her to recover these expenses, she therefore most humbly called upon and asked us to grant our imperial *Privilegium impressorium*[15] for a period of ten years to this treatise in quarto and by the power of it most graciously to deign to inhibit the worrisome damage caused by reprints. We graciously acknowledged the aforementioned right and proper request of the supplicant and thus granted her the favor and our imperial license and do this herewith by the power of this letter so that she may print the aforementioned treatise and reprint it, offer it for sale, and sell it, and so that no one can reprint, publish, or sell it without her consent and knowledge within a space of ten years, calculated from the date of this letter, within the Holy Roman Empire, as well as our hereditary kingdoms, principalities, and lands, neither in quarto nor in another format, nor with or without copper engravings or woodcuts. And we thus command each and every one of our and the empire's subjects, as well as the subjects and vassals of our hereditary kingdoms, principalities, and lands, especially, however, all printers, book peddlers, bookbinders, and book sellers, if they are to avoid a fine of a full three marks worth of gold, which each person, each time he transgresses our decree, will be subject to paying without fail—half of it

13. Leopold I (1640–1705, Holy Roman Emperor, 1658–1705).

14. Licensed by having sworn an oath.

15. Latin: printing privilege.

into our imperial treasury and the other half to the aforementioned suppli-cant Justina Siegemund. We do herewith fervently command and desire that you not—and no one on your behalf—reprint, distribute, offer for sale, peddle, or sell the aforementioned treatise on difficult and unnatural births within the designated ten years, neither in quarto nor other formats, nor with or without copper engravings or woodcuts, and that you in no wise abet others in reprinting, distributing, marketing, peddling or selling it. To avoid our imperial disfavor and the loss of that same favor for your printing, we command you to stand by the aforementioned Justina Siegemund with the help and cooperation of the pertinent authority of the area whenever you shall discover such things and immediately on your own authority without obstruction to confiscate everything and thus to act on her behalf. However, the oft-mentioned Justina Siegemund shall be obliged, at the risk of the loss of our Imperial License, to deliver at her own expense the customary copies of said treatise to our Imperial Court Chancery and to have printed and in-clude this *Impressorium* in it to inform and warn others. In witness thereof this letter is sealed with our imperial privy seal. Granted in our city, Vienna, on the fifth of May, in the year 1689, of our reign, in the thirty-first year of the Roman Empire, in the thirty-fourth year of the Kingdom of Hungary, and in the thirty-third year of the Kingdom of Bohemia.[16]

Leopold
(*L. S.*)
Vt. Sebastian Wunibaldt,
Hereditary Lord High Steward, Count of Zeyhl:
Ad Mandatum Sac. Caes.
Majestatis proprium.[17]
Frantz Wilderich von Menßhengen.

We, Frederick III,[18] by the grace of God, Margrave of Brandenburg, the Lord High Chamberlain of the Holy Roman Empire and Elector in Prussia, Duke

16. The document refers here to the various years of Leopold's official reign as Holy Roman Emperor, King of Hungary, and King of Bohemia.

17. Latin: by the proper command of His Highness, the Holy [Roman] Emperor.

18. Frederick (1657–1713) had as Frederick III become Elector of Brandenburg in 1688 upon the death of his father, Frederick William; he crowned himself King in Prussia in 1701, and as Frederick I became the first king in the succession of Hohenzollern kings who ruled the newly created kingdom of Prussia. Siegemund refers in the title of her book to his electoral court. Sophie Charlotte, one of the dedicatees, was his wife (see above, note 3).

of Magdeburg, Jülich, Cleves, Berg, Stettin, Pomerania, of the Kashubians and Wends, also in Silesia, of Crossen and Schwiebus, Baron of Nuremberg, Prince of Halberstadt, Minden, and Cammin, Count of Hohenzollern, the Mark, and Ravensberg, Lord of Ravenstein and the lands Lauenburg and Bütow, etc. We hereby proclaim for ourselves, our heirs and descendants, margraves and electors in Brandenburg, etc., as well as to each and every person that our dear retainer Justina Sigismund [*sic*] humbly gave us to understand that she intended to publish a treatise that she had written under the name of the Court Midwife of the Electorate of Brandenburg. This treatise had been approved not only by the medical faculty of our University at Frankfurt on the Oder, but also by various Dutch physicians and doctors. She most obediently requested that we deign to grant her a privilege so that she would not be subject to any damage, detriment, or inconvenience by pirated editions or the sale thereof in our lands. Not only has the supplicant recorded various proofs of her dexterity and adroitness, but the medical faculty of our aforementioned university at Frankfurt on the Oder is of the opinion, according to the attestation provided, that the treatise will serve the public weal. Thus we hesitate all the less to grant the aforementioned petition. In our capacity as Elector and Sovereign, we do hereby privilege and favor said Justina Sigismund [*sic*], mentioned at the outset, as follows: that within the Electorate and Mark of Brandenburg, in the lands incorporated therein, and in our lands and provinces the aforementioned treatise cannot be reprinted for twenty years or imported, or distributed, sold neither secretly nor openly, nor sold off cheaply, nor should it be reprinted outside of our territory. Rather, by the power invested in the authority and the crown by virtue of this, our open letter, they, one and all, shall hereby be forbidden in every case as follows: the copies shall be confiscated, and they shall be fined 500 thalers, half of which shall be paid to our treasury and the other half to the petitioner, along with the copies. We and our descendants, Margraves and Electors of Brandenburg, etc., thus desire graciously to protect and preserve the oft-mentioned Justina Sigismund [*sic*] for that period of twenty years. Thus we graciously exhort all our governments in our electorate, our duchies, earldoms, and baronial lands to do this in our stead and to enforce our privilege appropriately and to punish those who act contrary to it as mentioned above. The petitioner shall, however, also be obliged to deliver at her own expense four copies of this treatise to our library as soon as it appears. Faithfully and without malice: Nevertheless, Us for Ours and all others according to their rights without prejudice. Documented with a signature by our own hand and an attached electoral feudal seal, granted at

Cölln on the Spree, April 22, after the birth of our Lord and Redeemer, in the year 1689.

Frederick
(*L. S.*)
Eberhard Danckelmann[19]

A NECESSARY PRELIMINARY ACCOUNT TO THE WELL-DISPOSED READER

It is generally the custom to provide books with a preface in which the reader gets information about one thing and another, making him all the more eager to read the book. This is why I wanted to put this preliminary account instead of a preface at the front of my simple, but well-founded conversation about difficult deliveries and their safe completion with God's blessing and a skilled hand, namely,

How I came to this profession and this art,[20] what moved me to write this manual and allow it to see the light of day, and what expectation and purpose I had in doing so.

To this end that the curious reader may be all the more eager to read it, but that those who are disposed to doubt and contradict may be somewhat convinced.

1.

Regarding the first point: how I came to this profession and this art. I can do nothing but give God the honor, God, who miraculously called me to it and richly blessed me in it: I have never been with child much less given birth to one, yet have had to attend difficult births. I want to make this clear from the start to the know-it-alls because they think a woman who herself has never been in labor could not write a thorough account of difficult deliveries and perilous labor, and thus fancy my manual is not well-founded and indeed

19. Eberhard Freiherr von Danckelman(n) (1643–1722), tutor and chief advisor of Frederick III of Brandenburg who later fell into disfavor.

20. The German reads *Wissenschaft.* In this handbook Siegemund uses the word *Wissenschaft,* which in modern German can mean science, knowledge, or skill, to emphasize a special knowledge or skill that midwives acquire through study and experience, a notion akin to our modern idea of expertise. English handbooks of midwifery from the period speak, by contrast, of the midwife's "art." I have translated *Wissenschaft* variously as "knowledge," "art," and "skill," or with a combination of these words. Siegemund does not use *Wissenschaft* to mean "science" in the modern sense of the word.

dare to convince others of this. But whosoever examines this reproach with his reason will easily divine that it stems either from ill will or irrationality. Consider that it is not necessary for a person to experience physically all of the cases in which he would provide advice and aid. How unreasonable it would be then to conclude of a physician that he could not provide good medicament and council in cases of grave illnesses because he had never suffered them himself. As if he had not, by the grace and blessing of God, been able to bring about successful cures, with his unflagging diligence, long experience of many diseases, and diverse practice (even if he had never suffered consumption, dropsy, fever, or other illnesses). How often a physician knows how to describe a patient's symptoms quite clearly! He of course does not feel them, but has identified them based on his own knowledge or on that of others. Is it not then equally possible for a midwife, even had she borne no child, to judge and advise by means of assiduous inquiry, manipulation, and reflection not merely as well as, but better than those who have borne children? How absurd it would be to conclude of a surgeon that he could not heal wounds, could not set broken legs or arms, could not remove a gangrenous limb, because he had not experienced these injuries physically. Indeed, he can do so when he has adequate knowledge of the injury and knows what to treat it with, in particular, when his manifold experience with similar injuries enables him to make the correct determination.

Personal observation and experience of such cases show us daily that even physicians who are seldom sick and surgeons who were never injured can, with God's grace, give advice about the illnesses of others and heal their wounds by means of their knowledge and experience. This should set aside all opposition and carping: even if God never called upon a midwife to bear children in her marriage, He could make her, with many years of experience, skilled and competent to attend difficult births. Do we not have the example of clever and judicious physicians and surgeons themselves intervening in difficult births by dint of their well-founded knowledge and experience and thus delivering the woman in labor? What then of that groundless reproach that those who have not themselves borne children cannot attend difficult births? I have had to lay this out clearly for the simpletons so they will not be confused by the silly talk of obstinate people.

Moreover, consider the following case: God has blessed a woman in her marriage ten, twelve times, and more and suffered her to bear children. Would this lead to the conclusion that this woman—if she otherwise had no knowledge or experience of doing so—would therefore be able to advise in cases of difficult births and to intervene manually? If she had had easy deliveries, how could she exercise judgment about difficult ones? For it is often the case that those who have easy labor cannot believe or comprehend the danger of difficult births. Rather they judge others by their own experience. Or if they experience hard and unnatural labor, they are not enlightened by their own pain; indeed, they are in less of a position to give advice. Often their pain is so intense, their throes so violent, that they do not know what is wrong with them, let alone learn something from it. Even if they have endured it themselves, they are not capable of turning a child or managing an unnatural birth, unless they be afterwards instructed in the art of manipulation. Thus I do not doubt that those who were previously inclined to doubt or to use this reproach to gainsay my manual have been sufficiently enlightened. And if they will not recognize this, I cannot help them, regardless of what they say. I know I can count on the support of all sensible physicians, who have variously fortified me in my profession and practice, indeed, of all Christian and unbiased readers: a midwife, even if she has never endured labor pains, can, by the grace of God, nevertheless assist and serve mothers in labor in the most difficult births if she employs diligent reflection and has many years' experience and especially if she assiduously calls upon God and tirelessly pursues her profession, and above all, if she is convinced, as am I, that God intends to call her to it by a special dispensation and that she must recognize and embrace it to honor Him. Thus I will recount to the reader how I was led to this art in certain stages.

My dear, departed father, Elias Dittrich, pastor in Rohnstock in the principality of Jauer, died on me early on.[21] My mother, who is now also deceased, brought me up by herself and educated me properly until the age of nineteen, when I was married to my husband who is still living. At that time he was a steward in the employ of the town of Vielguth in the principality of Bernstadt in Silesia.[22] At age twenty[23] I was thought by all of the midwives to be pregnant. And when the forty weeks calculated for me were up, I was to give birth, or make ready to do so. The midwife was of the opinion that the child was in the right posture. Because I did not know any better than what I was told, I was in labor for three days without being delivered. They called in one midwife after another until there were four of them. They agreed with the first one that the child was in the right posture (when, however, there was no child). So, in accordance with their view, I was to be tor-

21. Rohnstock of the former principality of Jauer (Polish: Jawor) lies south of Liegnitz in modern-day Poland.

22. Siegemund's husband, Christian, must have been employed by the widowed Anna Ursula von Bernstadt (d. January 1, 1658), who was at the time living on her dowager's estate at the castle of Vielgut (Pulz, *Das Hebammenleitungsbuch von Justina Siegemund,* 29).

23. That is, as she expresses it, in her twenty-first year.

tured and kept on the rack for a fortnight, and my soul would have been driven from me before I had brought forth a child. And so this was the midwife's cold comfort: I would have to die along with the child. She had learned her trade so well in the lay fashion as to want a child from me that I was not carrying. God had mercy on me and fortuitously sent a soldier's wife into the village where I lay. She was brought to me in my dire necessity by my husband and mother. And because she had better grounding and judgment in this profession, inasmuch as she was also a midwife, she made this diagnosis: there was no child but a clogging of the blood, and, moreover, a serious disease of the womb and a fallen womb.[24] Thereupon we called in a doctor of medicine who with God's blessing and good medicament brought me around.

This peril then, once endured, as I recount in detail in chapter 4 of this book, pp. 90–93,[25] was the first step toward my profession. I recovered and was eager to study books and summary sketches that I procured for myself on this subject in order to learn a thing or two about my condition. And this was my first notion of this art and of learning birthing practices. Without all this I never would have resolved to make a profession of it, that is, if God had not given me the opportunity and had not called me to it, as it were, without my consciously thinking about it, and if He had not also implanted in me a desire to serve others out of Christian charity and as a result of my memory of the misery I had endured.

So because I also conversed with the aforementioned midwives and they saw such books and illustrations of various childbirths in my possession and they thought I had gained the rudiments and understanding therein, it came to pass that one of these midwives bade me respond to the dire extremity of a peasant woman who was in labor (as can be read about in detail in chapter 4 of this book, pp. [90–96]). I was still young, but twenty-three years old, and aside from what I had read in books and what the illustrations had impressed upon me, I was untried. It was a dire situation; the poor peasant woman was already in her third day of labor. The midwife, who was in fact the sister-in-law of the woman in labor, did not know what to do: the little hand and half the arm had come forth from her belly. They begged me to try, and I allowed myself to be persuaded for the love of my neighbor and intervened manually—as in the first copper engraving and the particulars on pp. [91, 94–96] in chapter 4. God was gracious, even more so than I re-

24. Prolapsed uterus.

25. I have here and elsewhere substituted the corresponding page numbers of the present translation. The inclusion of such cross-references is remarkable, for these references had to be added once the type had been set.

alized at the time, given my then limited understanding of unnatural births. So the child, though weak, was born and the mother delivered.

So I felt all the greater desire, like the drive of a divine calling, to reflect and think even more assiduously, especially because said midwife urged me to do it and took me along with her to many a woman in labor. I thereby gained greater and diverse experience, especially among the village poor. Whenever a danger presented itself, whenever there were difficult births, I was summoned from one place to another. Since the children were often already dead, it was my task, with God's grace, to save the mothers. I practiced in such a school for a good twelve years, while my husband was a steward on the Wartenberg estate,[26] and when one after another heard about me, I was fetched a good four, six, and eight miles, but for nothing but difficult cases and only to attend peasant women. I could never refuse, though I could never hope for remuneration. Rather, seeing that God blessed my work was reward enough: serving others and gaining more experience and acquiring the fundamentals in this art. Even after I had endured over twelve years' apprenticeship among poor peasant women, attending many a difficult birth and getting to know the wrong postures of children, even after I learned how to turn an infant and how to prevent danger, I was still not of a mind to make a profession of this work, seeing as how I was otherwise provided for by my husband and I only practiced this profession, as mentioned above, in cases of difficult births and when asked to do so. But it came to pass that I was summoned in difficult cases to attend some pastors' wives and finally to noblewomen and thus became known to physicians. One of them, who was in the municipal council of Liegnitz,[27] saw to it, without my knowledge or doing, that they requested and got me for their town midwife. So I spent some years seeking to serve God and my neighbor in this profession, and all the while I daily noted how one day taught the next and how God showed me ever-greater light in my profession.

A strange case occurred that provided me with the occasion for deep reflection and with considerable foundation in this art. It was a highborn person who had a growth in her womb that had started to rot and who faced certain death if she did not have it removed.[28] I would not have dared to

26. Estate belonging to a peer of the Holy Roman Empire in Lower Silesia.

27. Polish: Legnica. City in western Lower Silesia, now in Poland.

28. Luise of Anhalt-Dessau (1631–80), widow of Duke Christian of Liegnitz-Brieg-Wohlau (1618–72). When the duke died in February 1672, leaving only one underage son, Georg Wilhelm (1660–75), Luise ruled during the years 1672–75 as regent of the duchy. When Georg Wilhelm, the last of the Silesian Piast rulers, died three years later on November 21, 1675, Leopold I (1640–1705) laid claim to the territory. Siegemund claims that her highborn patient lived nine more years, which would place the event in 1671, but Friedrich Lucae's chronicle of

present this here (since as my Lord God knows I give Him honor in all things) if others had not thought it useful information and advised me to show it in the present copper engraving. Said highborn person was deathly ill and was experiencing dire symptoms. So one of the physicians got the idea that a moon calf[29] might be present. Various midwives were sought out; I, however, was taken aside and given some light in the matter by said physician and directed to try and see whether I could find anything: so I found a growth on the right side of the inward mouth.[30] The only solution was to remove it, but the question was how. As the picture indicates, I tried to get it with a crotchet,[31] thinking I could pull it out gradually; I discovered, however, that it was attached. God gave me an idea: I took a white ribbon, made a loop with it and with the index finger of my right hand (as can be seen in the copper engraving) placed the loop around the growth. When I had gotten a good hold on the growth, I pulled on the loop with my left hand and thereafter cut off the growth with a long pair of scissors so felicitously that this highborn person lived for nine more years. The three copper engravings show how this went: the first, hooking it and placing the loop around it; the second, tightening the loop; the third, cutting it with the scissors whereby the looped cord had been wound round my two last fingers so the growth could not slip back while I was cutting it. I not only became more enlightened in my art through this felicitous operation, but also came more into demand in various places to such an extent that I was released by the princely sovereign authority of the town of Liegnitz to take up a proper post and engaged to follow the court. In this profession I was summoned now to Saxony, now to Silesia, and was employed by many people. I thereby had the opportunity to reflect still more upon my profession and to learn from various learned physicians.

Finally, God arranged it so, to please a certain lady, I came here to attend her in her imminent childbed and thereby had the opportunity of being recommended to the late Elector of most glorious memory, Frederick William,[32] and was most graciously made Court Midwife by the same. The

---

Silesia from 1689 reports on Luise's serious illness at the beginning of her regency (1672) (Pulz, *Das Hebammenleitungsbuch von Justina Siegemund,* 44).

29. That is, mole. A molar pregnancy results when a sperm penetrates an egg without chromosomes and genes.

30. Cervix.

31. Hook.

32. Frederick William (1620–88), known as "The Great Elector," ruled Brandenburg-Prussia 1640–88, father of Frederick III (see above, note 18).

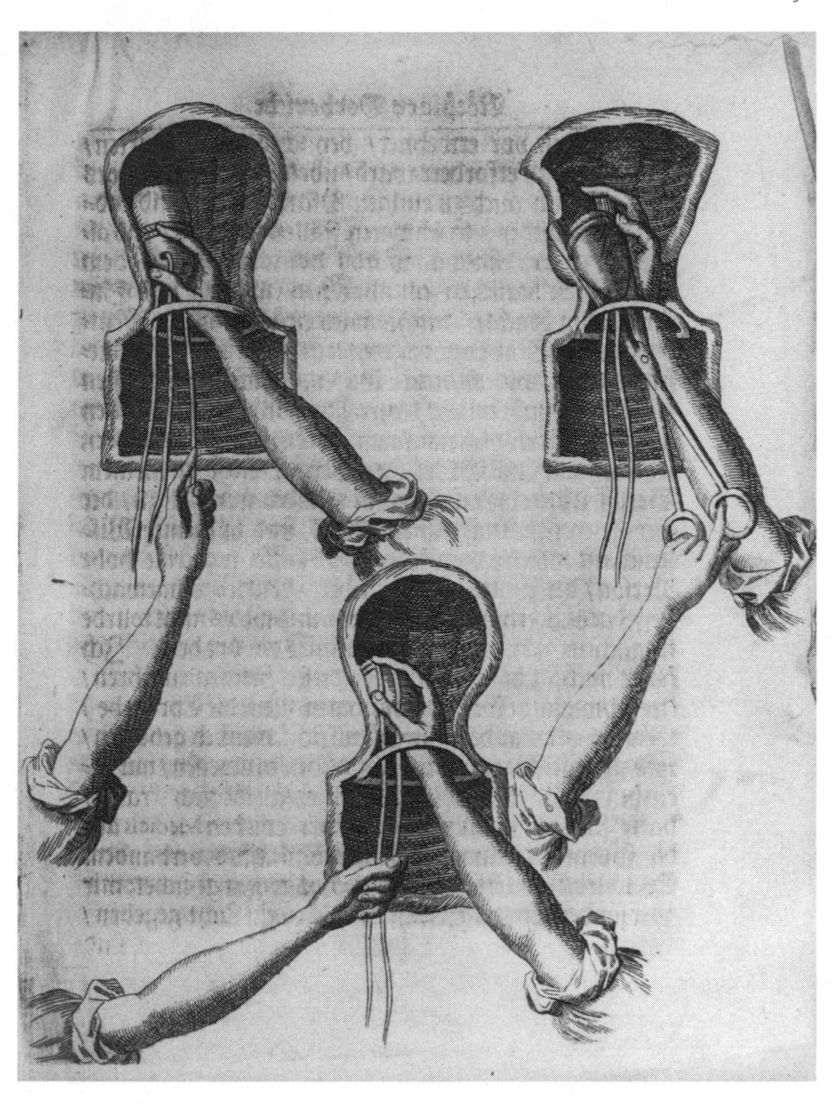

currently reigning Electoral Highness Frederick III has graciously confirmed me in this post. Thus God called me and to this day He has blessed me so I could serve my neighbor, not only highborn and illustrious persons, whom I could name here, but also the most humble folk. Especially when I saw them in danger and was summoned, I did everything in my power to advise and save them. This is thus the first thing I wanted to tell the Christian reader: how I came to this profession and art. Although I had no child, I nevertheless served many people in critical cases and helped deliver or save their children. God called me to it. Praised and thanked be He.

2.

But for me to recount briefly the other reason why I wrote this book and had it published, let this be known: because for many years I had to travel to various places and remain there until these women delivered, I found no better pastime than to put down on paper my thoughts and notes on difficult cases. It was easy for me to get a quill and paper, and I had in my head a supply of observations from my experience. So when I had an idle moment, I wrote down something without thinking that this should be a book for the world. Instead I wrote up this and that complication for myself to keep from forgetting them and so I could speak with others the more knowledgeably, especially because I often heard both midwives and wise women speaking in such an unfounded manner of various complications. It also happened that in a certain matter, which I present in its place in this book, I had to get testimonials and a verdict from eminent physicians in order to defend myself. I thereby had occasion to recall the difficult births from the previous years' experience. In writing, one question grew out of the other (and I see no end to them) so I finally resolved to present everything in a conversation that was expanded and improved on from time to time. I also talked about it with many highly learned physicians and variously read parts of it to noblewomen who often urged me to publish it. I could never really resolve to do it because first this and then that kept me from it until a year ago when at the command of His Electoral Serene Highness, my gracious Lord,[33] to serve Her Highness, the Princess of Nassau, in childbed I had to travel to Friesland and from there on to The Hague in Holland.[34] While I tarried there for a time both before and after the happy confinement, I was graciously allowed to show Her Majesty, the reigning queen of England,[35] the manuscript I had put together and already furnished with a few illustrations. Her Majesty was not only pleased with it, but she urged me to prepare it for

33. Frederick III of Brandenburg (see above, note 18).

34. That is, Henriette Amalie, consort of Heinrich Casimir II von Nassau-Dietz, Statthalter of Friesland (see above, note 7). Siegemund refers here to the birth of Henriette Amalie's second daughter, Maria Amalia, in 's-Gravenhage (that is, The Hague) on January 29, 1689. Friesland and Holland are both provinces of the present-day Netherlands.

35. Mary II, Queen of England (1689–94) (see above, note 5).

printing without delay. Moreover, because various doctors, especially the physicians-in-ordinary to the princely court of Nassau, advised me to do precisely this and offered me their assistance, I made this decision and seized the opportunity to specify the copper engravings I still needed and to have them made and thus to prepare the work for printing. I then traveled to Frankfurt on the Oder after I had completed my trip to Holland in order to submit this, my plan and book, to the highly judicious review of the medical faculty, who showed themselves willing to vet it and who, after reading my book, urged me to publish it. Thus this book, which was long in seeing the light of day, as if in childbirth, will be what I leave to the world, since I have borne no children. Therefore I need not justify at length my reasons for publishing it. Everyone is obliged to employ his gifts and knowledge for the good of his neighbor because we are joined to one another like members of a body. I cannot serve my neighbor better than by revealing in print the knowledge and experience I have acquired over many years, with the heartfelt wish that it may, by the grace of God, be applied with benefit wherever necessary, and especially since my summoning by the deceased electoral authority[36] of these parts was for the purpose of aiding other midwives in difficult cases and of helping others to this art, a calling to which I am still bound. But in my experience few people ask me for instruction except when they summon me when the laboring woman is in dire extremity. I thought I could perform my duty in no better way than to have a clear manual printed, which is what this conversation is. It is not just that midwives can learn from it, but it can be used to find out what foundation they have in their art, and they can be taught how, with God's grace, to take care to prevent many a bad complication. Finally, because my desire for our Eternal God increases with each passing day and my end is in sight, I did not want to hide my light under a bushel, that is, my God-given art and my experience of thirty years in this profession, but to honor God and to profit the common good. I had the pictures of the postures of birth engraved and printed out of love of my neighbor at my own expense, so I could repay the world with love and after my death could leave to the world the enlightenment in this art and experience that God gave me in this world. These are my reasons for publication. If anyone tries to impute others to me, he will find himself mistaken. And I will deign to give him no other answer but that God knows our hearts: nothing but the aforementioned command and my profession led me to publish this, nothing but the desire to serve my neighbor, trusting that perhaps those who now hesitate to ask me orally will be instructed through this writ-

36. Frederick William of Brandenburg, the Great Elector (see above, note 32).

ten work and will have a better grounding to serve usefully in difficult births. May God graciously grant it.

3.

It is, third, my sole purpose in this work and with these expenditures to serve the common weal, and especially to enlighten midwives so much so that I would like to pass on everything I know and thereby help suffering women in their travail. I will not complain here of the gross ignorance one perceives in many midwives to the grave danger of those who suffer under their care. There are many who have never gotten a solid grounding in it. There are others who are somewhat headstrong. It is highly necessary for all of them to read this manual and to learn from it what they know and do not yet know and every day to reflect more and more, and it is particularly necessary for them to get a good understanding of deliveries and of the postures of the infant right at the beginning and for them to impress upon themselves thoroughly according to the illustrations how children lie in the womb and in what manner they present and how they can figure this out and on what basis. Though there is instruction in this throughout the book, I wanted to make a deeper impression by introducing in this preliminary account a copper engraving of a full-term infant, showing how it tends to lie in the womb, now this way, now that way, but always curled up, the better to impress it upon midwives who as yet have not learned the rudiments and cannot grasp what they see in the following copper engraving, which shows only the womb and the children [*sic*] lying in it. So they should contemplate this large copper engraving well.[37] It illustrates by means of the following alphabetized letters and their captions how a teeming belly, along with the child lying in it and its cake with all the skins that belong to it, is constituted, according to which midwives can and ought to orient themselves if they mean to practice their art with care and in a Christian manner.

A. Cake or afterbirth[38]
B. The chorion, or choroid skin
C. The parts of the urinary membrane
D. The womb
E. The covers of the belly

37. This is a copy of table 56 in Govard Bidloo's *Anatomia humani corporis* (1685). My thanks to Lilla Vekerdy, Washington University School of Medicine Library, for enabling this identification.

38. Cake: placenta; skins: membranes.

F. The child
G. The amnion, or the skin of the sac or translucent membrane, which is otherwise called the caul.
H. The umbilical cord or navel string
I. The child's head
K.[39] The elbow
L. The knee
M. The hands
N. The chest
O. The neck
P. The feet

For it is not sufficient for a midwife to say she has attended many difficult births; it is better when she knows how to prevent difficult births so no one labors for two or three days. I have encountered various ignorant midwives who tell of difficult births that, after protracted labor, they managed to bring to a felicitous conclusion for the mother and sometimes for the child as well. And they succeed without any understanding or knowledge: as the proverb says, even a blind chicken finds a few grains of corn now and then. It cannot possibly be otherwise because they do not know what actually occurred over the course of the birth. To be sure they say, we pushed the child's little hand back inside every time it came forth until Our Dear Lord aided us and the appointed hour for the birth arrived. They say it went well. But I want to know how these children actually presented when they were born. Was the arm by the head and the fetus very small? They said no then. Thereupon I inquired further, did its feet emerge together with its little hands? Again they said no. Then I asked them further, did the rump present, and was the child thus born doubled up? They just kept on saying no. So we can see that they were sorely in need of instruction, namely, children, when they are full term and of proper size to be viable, can be born no other way than the above-mentioned three ways. To be sure one is larger or smaller than the other, but with a full-term infant, even the smallest one cannot be born in any manner but these three unless it be premature, as with abortion[40] or miscarriages at twenty weeks or more.[41] These can be clumped

39. The text skips from I to K here, conflating I and J, as was common in alphabetized dictionaries of the time.

40. That is, spontaneous abortions

41. Modern medicine distinguishes between "premature delivery" of a fetus "weighing between 500 and 2500 grams after twenty weeks and prior to thirty-seven weeks gestation" and "abortus," "an embryo weighing less than 500 grams . . . delivered before 20 weeks' gestation" (Ne-

together with the afterbirth, which a full-term child must leave behind. I spoke to them further then, telling them they really should know and would have to know in which of the three manners the children had presented and been born, because, as I said, they could be born in no other manner—

ville F. Hacker and J. George Moore, eds., *Essentials of Obstetrics and Gynecology* [Philadelphia: W. B. Saunders Company, 1986], 22).

through their own motion,[42] through protracted labor, or by being turned. But no better response followed, for they had no understanding of the vital turning of the child. Thus their so-called help was worthless, and Our Dear Lord, along with Nature, had to step in. When I persisted in probing the foundation of the knowledge they had at hand, they finally became angry and told me of crazy births, for example, that the children were born in a clump with head, hands, and feet all at the same time. Indeed, they said they presented with their rump and that their head was between their legs and they were born rolled up in a ball. Not to mention ridiculous accounts they gave of other births. Because I was angered at hearing such stupid things, I wanted to write this manual to help them along and to precede it with this copper engraving so midwives like that would realize they had not yet acquired sufficient knowledge and would be all the more desirous of researching the proper fundamentals. I would thus do well to repeat this one more time: a child can be born in no other way than in the above-described three ways unless an ignorant midwife should come along and quite irrationally rip off its arms, along with its ribs, lungs, liver, as well as its legs and even the intestines (I have, when summoned too late, witnessed such lamentable, if well-intended, help being administered). If the children are not too large, permitting one piece after the other to be torn off, the mother's life might be saved now and again. But she will hardly retain her health, and it will occur only seldom (or not at all) without the application of great force and injury to her. And even if it works with a strong woman and a small child, it is nevertheless terribly foolish and irresponsible. Should not such midwives be duty bound to fear God and learn better if they aspire to preserve a good conscience and conduct their profession in a Christian manner? As long as they do not know, see, hear, or read a better foundation in the art, we may be able to excuse them somewhat. But when they have been enlightened and still grope in the dark out of a hard, perverse mind-set or do not take up well-intended instruction in the basics of the matter with as much diligence as possible and do not serve their neighbor, this stubbornness will one day have serious consequences. I leave this to argumentative, careless midwives to ponder well. I truly hope they read my well-intended instruction dispassionately and soberly—this manual on difficult births and the possible avoidance of them, as well as the deft turning of infants in the wrong posture—and reflect thoroughly on what they have read and pay more attention when they are performing their services. And I hope when they do, they will applaud this manual—if not publicly with their mouths, neverthe-

42. In this period it was commonly believed that fetal motility aided the birth.

less with conviction in their hearts and minds—and will transform their previous thoughtless blind hostility in part into univocal amity and accept me into their company, even if I have had no children. For I have demonstrated that [giving birth oneself—*trans.*] can do little to help or hinder the acquisition of knowledge and skill, and that it is, on the contrary, obvious that there are many ignorant midwives who themselves have had many children and still are none the wiser for it.

God does not bind his gifts to particular kinds of people but gives them to whomever he wishes, and to whomsoever he gives these gifts He grants also the desire to research and reflect. Thus, in this way He blesses diligence with still more gifts and enlightenment. May the honor be His in all things if my neighbor can be edified by this work. I have completed it with good intention and moreover intended this preliminary report to provide greater enlightenment. I must, however, add one more thing before I close.

It is possible that when they hear about this book, some will speak against it, saying it does not issue from my brain; when others read it, however, they will consider it just too simple. Regarding the former, I can let others think what they want; the manner of the book will sufficiently show it is my work, for I am accustomed to speaking about things as I find them daily in my profession and I have recorded everything here as I speak of it. For this reason it will appear much too simple to some people it might in fact inform. Precisely because I wrote it as I saw fit, it cannot appear in any form but this one, and I believe I can thereby most readily achieve my purpose. In the beginning I assembled these questions for the sole purpose of enlightening a few friends since I had experienced so many cases in which mother and child came into extreme danger as a result of a small mistake (that could easily have been prevented). They urged me not only to illustrate graphically how a child can be guided to a natural birth, or if it be in the wrong posture, how it can be turned, but also to provide a clear description of this and more and thereby to enlighten midwives.

And to this end it was necessary to present as clearly as possible these complications that are so little known and to repeat some things often so even the most simple people might grasp them. Learned things are appropriate to the learned, who can see in my instruction only this: how carefully touching a woman[43] can divine the danger and how, with God's mercy, careful manipulation can serve and help in a dangerous situation. If not everything be written as learned men would have it, I will gladly let myself be in-

43. Palpating the cervix.

structed, and if one thing and another can be explained still better, I will always show myself ready to do so, insofar as I can speak of it from my understanding and experience. Where I am mistaken, I will thank him who with well-founded knowledge teaches me better; should I be given additional cases not reported on here, I would be pleased to learn about them to increase my knowledge, for we can and must become wiser through experience day by day. If my purpose and my person are considered rightly, since I seek to instruct others as a midwife, no one will expect from me any kind of writing other than this clear and simple one, and for this reason no one will disdain the necessary and useful things presented herein. The testimonial of the praiseworthy medical faculty at Frankfurt states that many good things inhere in this simple manner of speaking. They read it before I had it printed, and after reading it, the venerable electoral court chaplains testified that there was nothing superstitious or contrary to God's word therein. I had to get their testimonial to get the privileges.[44] Both of these testimonials herewith inform the reader how this, my simple work, insofar as is necessary, can be justified. My preliminary account will be appropriately ended with them.

Honoring God with His gifts and serving my neighbor through my profession is my purpose in this work and will also be my aim as long as God gives me strength and life. I commend myself and the reader to His mercy with the wish that God prepare us all in this life for the purpose of His praise and everlasting bliss. Amen.

We, the undersigned do hereby acknowledge and attest that in reading this book on difficult births and preventing them when possible, including turning children in unnatural births, which is presented in the form of questions and answers by Justine Siegmund [*sic*], sworn midwife of the court of the Electorate of Brandenburg and which was turned over to us for proper review, we found nothing that disputes God and His holy word and nothing in the least injurious to the Christian faith. Rather, we found everything to be disposed with propriety and free from superstition. Thus her Christian intention is rightly to be praised and this, her manual, can right well and

44. The three privileges that Siegemund procured are the best guards against pirated copies that early modern Germany had to offer. Nothing like modern-day copyright existed at the time.

beneficially be printed for the good of the public. Sworn at Cölln on the Spree, April 5, 1689.

G. C. Bergius. D.
Elder of the Electorate of Brandenburg
Court Chaplain and Consistorial Councillor

Heinrich Schmettow
Electoral Court Chaplain

Antonius Brunsenius
Electoral Court Chaplain[45]

We, the Dean, Senior, Doctors, and Ordinary Professors of the Faculty of Medicine at the University of the Electorate of Brandenburg in Frankfurt on the Oder hereby let it be known that Frau Justina Sigismund [*sic*] gave us a book to read consisting of two parts whose titles are (1) A Conversation between Two Peace-Loving Midwives, who converse frankly with one another about their profession or office in the matter of difficult births, as well as on the subject of turning children in unnatural births, the interlocutors being Christina and Justina; (2) A Second Conversation, wherein Justina learns in turn from Christina by asking many questions whether the latter has understood and comprehended her instruction.[46] With this appended report we testify that we are not disposed to let ourselves be put off by its simple style since she wrote it up on her own without the help of others particularly so that simple midwives could comprehend all the more easily the things contained therein and could thereby educate themselves for the common good of pregnant women and women in travail.

Inasmuch as we approve of this, her intention, we have also found upon reading through it that in this manner of writing she has named and described from her own experience so many beneficial and useful things, adroit manipulations, and ways of turning the child, of which unfortunately

45. Georg Konrad Bergius (1623–91), Heinrich von Schmettau (1628–1704), and Anton Brunsenius (1641–93) were, respectively, the eighth, ninth, and eleventh court chaplains at the Brandenburg court in Berlin-Cölln (Rudolf von Thadden, *Die brandenburgisch-preussischen Hofprediger im 17. und 18. Jahrhundert. Ein Beitrag zur Geschichte der absolutistischen Staatsgesellschaft in Brandenburg-Preussen*. Arbeiten zur Kirchengeschichte 32 [Berlin: Walter de Gruyter, 1959], 170).

46. The testimonial does not in fact quote verbatim the titles of the two parts, but rather describes here the format and contents.

up to now most midwives have been ignorant to no small disadvantage of many expectant women. Therefore we are of the opinion and are hopeful that it will be published with benefit, be read by midwives, and be received by all women with profuse gratitude. May the Almighty fulfill our hope and, as He has up to now, bless her in her profession in accordance with her own wish.

Given under the seal of our faculty, Frankfurt on the Oder, March 28, *Anno*[47] 1689.

Dean, Senior, Doctors, and Professors of the Medical Faculty at the University of the Electorate of Brandenburg at Frankfurt on the Oder

(*L. S.*)

47. Latin: in the year.

PART ONE
CONSISTING OF HIGHLY NECESSARY INSTRUCTION IN DIFFICULT BIRTHS AND THE AVOIDANCE THEREOF WHERE POSSIBLE, AS WELL AS INSTRUCTION IN DEFT TURNING OF CHILDREN WHO LIE WRONG

INTRODUCTORY CONVERSATION BETWEEN TWO PEACE-LOVING MIDWIVES

CHRISTINA: Dear sister, pray, give me a thorough account of difficult births and the avoidance thereof where possible, as well as of deft turning of children who lie wrong. Can a midwife help laboring women in these unnatural cases, which occur so often?

JUSTINA: Yes, she can help in certain ways, namely, if she is intelligent, has attended sundry births, and has a skilled hand.

CHRISTINA: I know you have many years' experience, so I ask you once again to share your knowledge with me, to honor God and benefit your neighbor.

JUSTINA: Very gladly. Just show me what you desire and what you are especially eager to know. Have you served a lot of women in their travail?

CHRISTINA: Something like two hundred.

JUSTINA: Then you must already have a good foundation in the art. Tell me then, did you always have natural births[48] with all these women?

CHRISTINA: No, I had a lot of unnatural births among them. These were difficult and dangerous.

JUSTINA: Tell me then, about the unnatural births. Did the children present with their hands or feet? And how was it when the pains kept on coming as they do when the birth is commencing and you have to touch her to see whether it is true labor and the child is lying right?

48. By "natural birth" Siegemund means simply that the infant is born with the head first.

CHRISTINA: That is certainly a strange question! Who can tell by touching early on when the pains commence how a child lies if the waters have not yet broken? The child is still high up in the belly, and you cannot reach it until the waters break. Not until then is the birth at hand, and if the child lies wrong, then that very member of the child, be it the hand or the foot, often even the navel string, comes forth before the child does.

JUSTINA: Say the hand or the navel string comes forth with the waters. How does the child lie in the womb?

CHRISTINA: It lies wrong. When it lies right, nothing can come forth. When the head comes forth, then everything else comes right out, but if the hand or the navel string or the feet come forth, then the birth will often take a long time, though sometimes not that long.

JUSTINA: Why does it take one woman longer than another to give birth if the children, as you say, all lie wrong in like manner?

CHRISTINA: Who can know that without God's help? Who can do anything? You have to wait for the proper moment when God will aid you.

JUSTINA: But do you not intervene if the child presents with a hand or the navel string? How do you assist?

CHRISTINA: What else can I do but push the hand and the navel string as far back inside as I can manage and then hold them back until the child is born?

JUSTINA: If the child cannot be delivered within a few days, do you keep on holding back its arm?

CHRISTINA: Yes, if the child is dead and it takes too long, then you have to let things take their course. Who can help in this case?

JUSTINA: Know you no other way of assisting besides putting the arm or the navel string back inside and holding them in?

CHRISTINA: I must in any case see to it that I provide assistance if it comes forth sooner, and then Our Dear Lord will help too.

JUSTINA: I think it will work if God wills it. But I want to know what you do and can do as necessary?

CHRISTINA: I feel my way up the arm and push the child as best I can in every way possible so it can be born at last.

JUSTINA: Can it be budged if the waters have already broken and run off, and are children's bodies always in the same posture for birth if their little hands come forth first?

CHRISTINA: Who can really know all this? Our Dear Lord helps when it pleases Him. I do as much as I can, push and help the child in all manner of ways until it finally comes forth.

JUSTINA: I hear and see you manage it blindly. You are still lacking in knowledge of the matter. If you desire it of me, I shall teach you more about it, namely, as I said in my preliminary account, children can be born in none other than these three ways, either naturally or by being turned: (1) head first, naturally, (2) both feet first, and (3) rump first. If I were to quiz you further about how the children came forth when they were born, you would come up with the same thing if you thought about it, because it cannot occur in any other way, and when children cannot be forced into those same postures with strong pains and strong women while it is still possible, mother and child will lose their lives if there is no midwife on hand who can turn the child. Those midwives who know how to turn children properly do not allow it to become so perilous, as I shall further report at your desire and request. Above all, you must inform yourself about the womb and its parts so you know it thoroughly, along with its inward mouth. Thereafter, everything will take its due course.

CHAPTER 1

Of the womb and how to fathom it, what the inward mouth is, and whether it is necessary to know about all this

CHRISTINA: Tell me, then, dear sister, whether all women have a womb.

JUSTINA: Yes! All women must have a womb if they are to conceive and bear children.

CHRISTINA: Are there not more generative parts than just the womb?

JUSTINA: As far as the outward parts are concerned, I know only of the sheath or neck[49] that leads me to the inward mouth.

CHRISTINA: What is the inward mouth? How is it to be sought or fathomed?

JUSTINA: The inward mouth is the lock of the inward belly, that is, the womb, where the child is conceived, carried, and kept until natural childbirth takes place unless the pregnant woman should do violence to it or, as often happens, something violent happens to it, causing the child to be aborted. So in such cases you can be guided by the state of the mouth. A woman can labor all she wants, but as long as the mouth of the womb does not open up, the child can easily be kept inside and the pains stopped.

49. Siegemund uses the words "sheath" and "neck" for the vagina, as was common in seventeenth-century English as well. She, however, never uses the Latin "vagina"—unlike her university-trained German contemporaries and, indeed, unlike her English contemporary Jane Sharp, who uses the Latin derivative "vagina," as well as "sheath" and "neck." In the testimonials included in chapter 8, the German doctors use the Latin "vagina."

Then you must seek the counsel of physicians. If a doctor is lacking, there are tried and true remedies you can use. If the child penetrates or presses against the mouth so it opens from pang to pang, then nothing will keep the fetus inside. Thus you can make a diagnosis according to the condition of the mouth. It is the same with the mouth in full-term births, because in true labor it opens wider from pang to pang. If, however, it is false labor, the mouth tends to close up rather than to part. You will have to take note of the many differences associated with false and true labor. It is not possible to describe or consider all of them here. If you know nothing about the inward entrance to the womb, you cannot speak with foundation of any difference, neither of full-term nor premature births or of lying in the right or wrong posture, and you simply have to wait and see what comes. But by the time the child pushes into the birth passage, a great many mistakes and omissions have occurred. It is best to turn it when there is room; I am of the firm opinion that much pain and misfortune can thereby be prevented.

CHRISTINA: If I were to find the inward mouth right away and examine it, how would I prevent the complications that so often occur, especially with a mouth that is shut tight and does not open until the birth? What if a woman has been in labor for a whole day and still I do not feel it opening up at all and then violent pains ensue and force nearly everything into the birth passage all at once?

JUSTINA: It is not necessary to do anything if the mouth is closed. So you need not answer to God or man for this; no person can do anything in the dark. On the other hand, you need to be aware that I have never encountered a closed mouth with true labor. So you cannot even call it labor if the mouth is closed when the very first pains commence, unless it be a woman of advanced age who has never given birth and whose belly will not leave off carrying the child. This occurs only rarely. This circumstance cannot be prevented, and such people generally have hard labor. They can be aided, but I do not know how to make it understandable to you on account of your lack of experience; it might do more harm than good. Nevertheless, you need to pay attention, that is, to the mouth. You will discover how many dozen pains result in a small opening in such difficult births, and you will be able to provide comfort from one hour to the next by telling the woman in labor how much the mouth has changed.[50] You will need to hold back the mouth as soon as there is an opening large enough for you to introduce two fingers because with such difficult births a falling of the womb[51] can easily

50. Dilatation and effacement of the cervix.

51. Prolapse of the uterus.

ensue. You cannot be too careful of it, and it is better to hold it back (I speak not of pushing it back hard, but of holding back the inward mouth so it is not injured by the throes coursing through it), for I believe if you hold back the mouth, no injury will result. But if it is pushed back violently, the fetus will be held up or the mouth will be likely to be injured. In either case it would be indefensible. Holding back the mouth when necessary averts the danger of the sheath and inward mouth falling[52] and promotes the birth because the pains can course through it and are not inhibited. I wish to make mention here of the shifting or advancing of the inward mouth as well. I call it advancing before the head. I will describe it to you so you can understand it better. Most midwives who understand something has gotten in front of or lies in front of the child have a habit of saying the bear-mother (as they call it) has lain down in front of it.[53] This generally applies to bellies wherein the mouth lies deep inside and closer to the rectum than straight down. Thus they touch what lies before the child but do not know that the womb must have an opening, that is, a mouth. And they are as inexperienced as you are and think things will suddenly get moving after a lot of fomentations and fumigation. And it happens, but in no other wise than with the throes, and so they have no idea what happened.

You, however, should seek the mouth of the womb soon after labor commences, as I instructed you, and you will soon become thoroughly acquainted with this complication, know what it is, and how to help with it, namely, a gentle pushing against the child's head facilitates birth, and you can thereby prevent a fallen womb. If the child remains lying there until the throes force it out, it can easily result in injury. For I myself have gotten to know such people of whom there are now and again more than enough. This predicament occurs especially in such cases when the labor pains are pushed hard and the midwife screams with all her might for help. It would be desirable for all midwives to understand the workings of the inward mouth; many women's bodies would thus remain hale and hearty, and the midwife would not come under suspicion of having been the cause of injury. They are at fault, but unknowingly, because they know not what they do. The violence of the throes and the child lying in the advanced mouth of the womb tears everything from the belly, and the woman is not aware of it, nor is the mid-

52. First-degree and second-degree prolapse, the former being when the cervix remains within the vagina and the latter when the cervix protrudes beyond the introitus.

53. In other words, over the course of the infant's descent through the pelvis the cervix has descended below the pelvis, with the os pointing toward the coccyx. When the midwife performs a manual examination, her hand runs up first against the flesh of the uterus and she does not reach the os. I thank Amy Ravin, M.D., for explaining this condition to me.

wife, until the woman tries to get up during the lying-in period.[54] And gradually it becomes more and more apparent and worse and worse. Such a condition seldom improves and usually worsens. To be sure, the same sort of thing happens if a woman gets up too soon during the lying-in period, lifts something heavy, carelessly slips, or falls violently during the first nine days. Thus a woman can easily forfeit her health in this and other ways if she does not take good care of herself. No midwife can answer for that. If during the birth you are careful of the woman and protect the mouth of the womb in good time, you can avoid this danger and grave liability with a good conscience. Likewise this mistake as well: a woman has been in labor for an entire day, and you think there is no opening. Suddenly a violent pang sets the birth in motion. Precisely this belief of yours is in error: you do not know how to find the mouth and therefore imagine in the case of such an advanced mouth that there is no opening because the opening is pushed high up against the rectum. If you learn how to find the mouth, you will tell a different story and be capable of averting many a danger. I shall give you a further lesson. After I have acquainted you with the womb and the inward mouth—to speak as women do—by means of natural touch, I shall show you here in a copper engraving how the womb, along with the ligaments, and the mouth look. Under the letter A, you see how it looks in nature—an important physician, Regnerus de Graaf,[55] explains it in this engraving. You can acquire a lot of information by carefully contemplating the nature of the womb. You can see it here as it lies against the part farthest to the back and thus you can see the inward mouth so you will not be confused and think to find it close up inside when you seek it. It lies farther back near the rectum as the letters in the copper engraving indicate. And that is enough on the subject of the mouth and closed, hard bellies, which (thank God) are not common.

Explanation of the Letters in the Engraving[56]

A. The wrinkled inner substance of the sheath
B. The protruding inward mouth in the upper part of the sheath

54. Siegemund's language is ambiguous here. She probably refers to a prolapse of the uterus, but it is also possible that she means that the pressure of the child's head and the cervix's pointing back toward the rectum caused the cervix to rupture.

55. That is, Reinier de Graaf.

56. This engraving is a copy of plate no. 7 in Reinier de Graaf's *De mulierum organis generationi inservientibus tractatus novus* (1672).

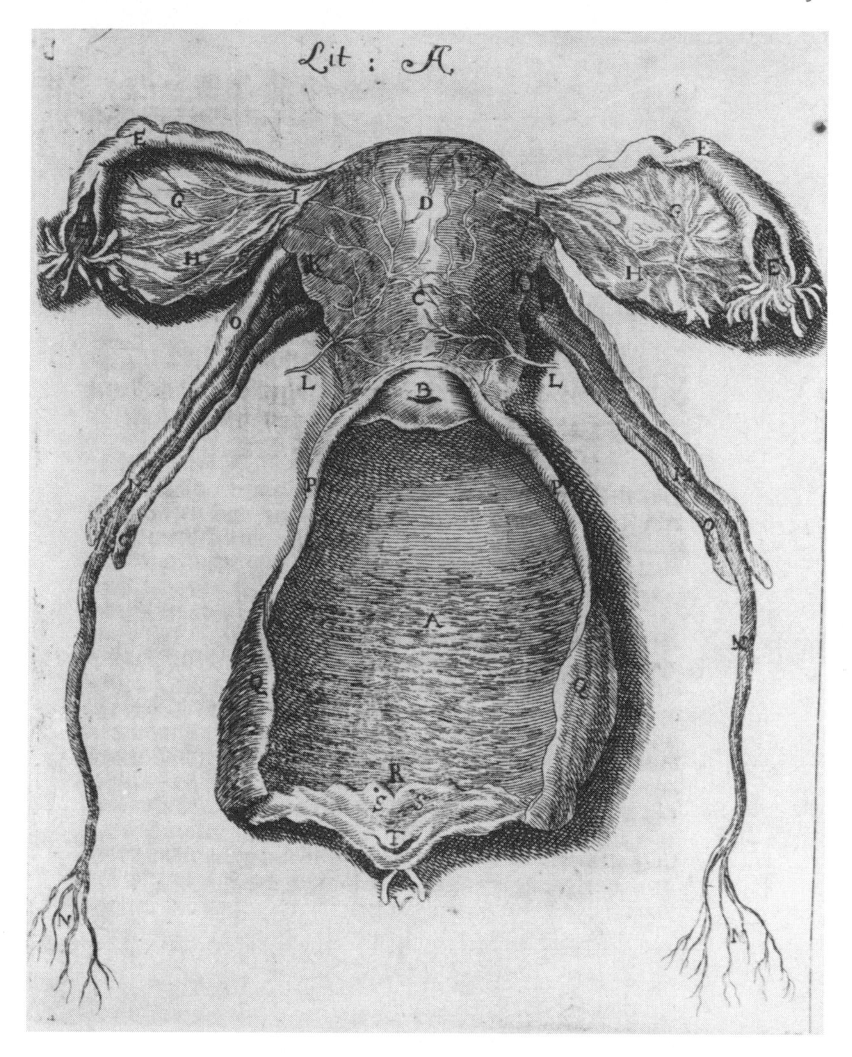

Figure A

C. The neck of the matrix[57]
D. The fundus of the matrix
EE. The horns of Fallopius or ejaculatory vessels[58]

57. The text here labels the cervix the *Mutter-Hals* (neck of the womb) in accordance with the original Latin label in De Graaf's work, *uteri collum.* Elsewhere, however, Siegemund usually refers to the vagina as the *Mutter-Hals.*

58. Oviducts or fallopian tubes.

FF. The inward fold or wrinkled matter of the ejaculatory vessels

GG. The nerves as they fan out through the wide ligaments of the womb, referred to as bat wings

HH. The stones[59]

II. The ligaments of the stones, commonly called the preparing vessels[60]

KK. Portions of the broad ligaments

LL. Little nerves that run back and forth through the matrix

MM. Portions of the round ligaments of the womb that are in the belly

NN. Portions of the round ligaments of the womb that are outside the belly

OO. Portions of the covering of the belly that encircle the ligaments as far as they go with them

PP. The thin upper part of the sheath, which is pulled asunder

QQ. The tough or thick part of the membranous substance in the lower part of the sheath

R. The outlet of the urinary passage[61]

SS. The folds that are to be found here and there in the glandular parts of the sheath[62]

TT. The woman's yard, or clitoris

VV. The large nerves that fan out over the upper part of the woman's yard (or clitoris)

CHRISTINA:[63] I asked you in the very beginning about the inward mouth that is closed up tight, and you answered, instructing me to pay attention to how many dozen pains would result in a small opening in a difficult birth, etc. But I should really have questioned you further: why is the inward mouth more yielding in one woman than in another, and what can you do about getting it to give way? And are there still other dangers besides lingering la-

59. Ovaries.

60. Ovarian vessels. De Graaf's term is *Vasa deferentia* (carrying vessels).

61. External urinary meatus.

62. The original Latin reads "Caeca lacunarum glandosum corpus perreptantium foramina," which H. D. Jocelyn and Brian Setchell translate as "the blind apertures of the lacunae traversing the 'prostatae'" (H. D. Jocelyn and Brian P. Setchell, trans., *Regnier de Graaf on the Human Reproductive Orders: An Annotated Translation of* Tractatus de virorum organis generationi inservientibus *(1688) and* De mulierum organis generationi inservientibus tractatus novus *(1672)* [Oxford: Blackwell Scientific, 1972], 175).

63. In the editions of 1690, 1708, and 1724 the conclusion of this chapter appeared at the end of chapter 7 with the following explanation: "The following three pages belong to the first chapter on the inward mouth." The error is corrected in the edition of 1723 and all following editions (except 1724). For the sake of clarity, I have, in keeping with later editions, moved these pages to their proper place at the end of chapter 1.

bor[64] to be feared with this condition? So I hope you will not take it amiss that I return at the end of this chapter to the questions I forgot to ask in the appropriate place.

JUSTINA: The natural cause of the inward mouth yielding sooner in one woman than in another is that the skin around the womb is more delicate with some and tougher with others. The more delicate skin gives way sooner than the tough one. You need to pay heed to this, for generally the most delicate women have the easiest births. By contrast the strongest women have the most difficult ones. And age plays quite a role here: if a woman is over thirty at the time of her first confinement, then the skin will be much tougher than in younger years. The older, the tougher. But with the old ones, there is the difference that the skin tends to be tougher with some than with others. If such a very tough skin is to give way, you need a lot of strong pains; with such great and violent pains, the inward mouth often gets a cramp that causes it to close up completely while the pains are occurring, as happens with false labor pains. So the pains accomplish nothing. Pains are themselves a kind of cramp. Now everyone knows that skillful rubbing, also grabbing or stamping and kicking, can help a lot with a cramp, that is, if the cramp is in your extremities. It is possible to do something like this with the inward mouth to prevent cramping, if not completely, nevertheless to a large extent. This closing up from cramp generally happens at the start of the birth, if the inward mouth cannot yield on its own, and it is difficult to distinguish from false labor. The inward mouth spontaneously and involuntarily stiffens and remains rigid until it is finally forced with violent pains or through the skillful intervention of the midwife. You must take care of this with the following intervention: you introduce two fingers into the inward mouth and hold them quite still until the pains come. You can tell if cramping is occurring; if so, the mouth closes up tight, even more so than in false labor. You can orient yourself accordingly and hold the mouth open as much as possible with your fingers. The cramp will thereby be eased and will cease. When intervening like this, you must also be guided by where the skin that encloses the child and the waters is pushing forward, since it is then quite easy to guide it into the inward mouth, and you have to arrange the position of the laboring woman accordingly. If the waters are guided right into the inward mouth, the opening can and must gradually give way and the cramping will also stop sooner than when these waters push to one side of the mouth (so you rightly understand it, I mean this water bladder that encloses the child and the waters). This situation is often very dangerous, since the

64. That is, obstructed labor.

child's head follows the waters. If you know neither how to guide the waters right nor how to manage the position, a cramp like that can set in. So even if there was no cramping at the start, cramping delays the birth for a day or more, even if no other injury ensues, though a falling of the womb can also occur and the woman has to suffer from it all the days of her life. Should violent pains force it on their own, these pains and the accompanying cramp are at fault; all this can, however, be prevented by the intervention and guiding in I have described here.

CHRISTINA: What kind of danger does it present when the mouth lies too far toward the rectum and is pulled up too high by the fetus in the case of bellies where the womb is tilted forward? Is it not possible to avert this danger?

JUSTINA: An inward mouth that lies toward the rectum will result in a very difficult birth if the midwife's two fingers are not inserted, as I already explained, when cramping occurs, because said water bladder and the child's head cannot engage and enter it. Particularly when the fetus draws the mouth upward in the case of wombs that tilt forward, the child has a tendency to get fixed on the sharebone[65] since the pains can slowly push the child on their own. Children that fix like that turn readily since they are up high and at this elevation have room to turn. Even if the waters and the child descend into the birth passage as they ought, the child does not encounter the inward mouth in the middle. So the pains cannot get hold of the mouth, and a very hard birth must follow unless assistance be provided; without the midwife's introduction, or rather insertion, of two fingers, the pains will only slowly and arduously force the mouth to open. The pains, together with the waters and the child, will form a sack as it were, even cause a falling of the womb, before the mouth will ever begin to open.[66] Even if the womb is sturdy enough not to fall, the birth will nevertheless be slow and hard. All this can be prevented by the insertion of those two fingers, and the birth can be made much easier.

CHAPTER 2

Of the pelvic bones, whether it is certain they must part over the course of labor and whether, if this happens slowly, a difficult birth ensues

CHRISTINA: Tell me what women's pelvic bones are like; you have talked here at length about the mouth of the womb, but you have not men-

65. Pubic bone.

66. In other words, the cervix has descended below the pelvis, pointing toward the coccyx.

tioned them. Most people are of the opinion that they have to separate in all births.

JUSTINA: You want my opinion on the pelvic bones, whether they must part when women are in labor. Well then, I will tell you I do not think much of this belief. But I leave everyone to their own thoughts on it. I do not want to give you any other instruction in this regard as you and I can be guided sufficiently by touching the inward mouth of the womb as to how much the belly has to give way during hard or easy labor, for the mouth can be reached during labor. Let us say the pelvic bones would have to part. Even so, it is not possible to know they do, because you cannot feel the pelvic bones on account of the flesh around the birth passage. Thus it is not necessary for you or me to know whether they separate because we cannot tell or get any information from it. Even if they had to part, the inward mouth orients itself to the pelvic bones, and we see the former yielding to the latter. We midwives can and must be guided by that, as I have already instructed you at length. My reason for believing that the pelvic bones cannot part is the following: I have been summoned to many difficult births, to aristocrats as well as others who had been in hard and difficult labor for three days and longer. They had not been spared by their midwives; in fact the midwives had managed the birth so foolishly that they had violently ripped off the children's arms and their legs too, which I found strewn about in the chamber—in their terror they had thrown them there and could not do anything more. And I could name those places where it happened upon demand. Indeed, I found they had torn the infant's ribs from its breastbone and cut their fingers on them. They had pulled violently on these broken-off ribs with pieces of cloth and still they could not help for they knew nothing about turning the child. Such cases and many more have come under my care. Yet Our Dear Lord has always enabled me to deliver these mothers of their children, and they joyfully thanked me for it. Various of them, and indeed most of them, survived, but some, who had suffered too long without proper aid, died afterwards. So were it possible for the cartilage between the hard pelvic bones to stretch and give way, how many people would have been injured during such difficult births by imprudent midwives? But I have never in all my days seen a fracture or a sprain of the pelvic bones or ever heard anyone complaining of such, whereas other injuries like the womb or sheath of the womb falling or slipping out of place can and do happen far too often as a result of difficult births.[67] To be sure, the nethermost bone,[68] that is, the last one on the backbone, can be injured and broken if the mid-

67. Prolapsed vagina or uterus, tilted uterus.

68. Coccyx.

wife is not careful of it. But that is not the pelvis or pelvic bones. Otherwise, most, if not all, women who have suffered difficult births and the administrations of inexperienced midwives would have had them torn asunder. I do not want, however, to quarrel with anyone on this account, but leave everyone to their own ideas about it. It is enough for you and me to know how to be guided by the mouth of the womb.

CHRISTINA: Dear sister, tell me whether it is correct and beneficial in the case of a difficult birth for the midwife to stretch the woman's belly while touching her or hold it open before and when the child appears and makes ready to pass through to keep the child from getting fixed if the birth is too hard and the child too large?

JUSTINA: If the birth is difficult and there is no other cause for its difficulty than the narrowness and tightness of the belly, that is, of the front part of the womb,[69] you have to give the woman time. You certainly should not stretch or dilate anything with your fingers. This is a common mistake. This sharp stretching injures the woman's belly and causes swelling before the child gets that far and comes forth. Thus the pain of the child's passing through is all the greater because of the swelling and the injured belly. I have observed that it hurts more than it helps. Proper aid must occur at the child's head, where it is jammed in the tightest, and not down in front where there is no child. As long as the child gets as far as the front part of the womb, then it will come down farther, even if it fixes there for a while. You have to give it time. If the head is not yet born, then it will hurt neither the child nor the mother. It more readily injures the woman's belly if you dilate or stretch it open so the birth passage tears. It is more necessary to protect the belly than to stretch it if the birth passage is not to tear, as often happens with the rectum as well, when the rectum is injured so it can no longer close up. This is a grave misfortune for the woman, for she cannot control her bowels at will thereafter.

CHRISTINA: Dear God, it is so dangerous when tearing occurs! It often happens with the first birth. It cannot be helped if the child gets stuck while passing through. How can you prevent it then if it is so dangerous for women?

JUSTINA: As far as tearing is concerned, there is a big difference between violent tearing and what occurs in first births, when it of course is not

69. Siegemund uses one of the contemporary terms for lap or womb, that is, "Schoß," rather loosely here to indicate the vagina, as was not uncommon in English handbooks from the period as well, where "womb" can include the vagina (cf. Jane Sharp, *The Midwives Book, or the Whole Art of Midwifry Discovered,* ed. Elaine Hobby, Women Writers in English 1350–1850 [New York: Oxford University Press, 1999]). Likewise she uses the German word for belly loosely here when speaking of the vagina and the perineum.

always possible to prevent a little tearing. But such violent tearing can be prevented if you leave off unnecessary and clumsy stretching or pulling open, especially when the child appears. It opens on its own, and the belly would burst if God's goodness were not so great. So you can imagine what happens if the midwife applies still more force with her fingers and dares to break open the belly. In such cases the woman's belly would have to tear even if it were not a matter of a first birth where it often occurs. Take care not to do that; it is completely irresponsible. Nature will help her on its own, unless the child be pushed up too hard against the rectum or is about to be. In that case you can simply push up the head a little, and it will lift itself out and go straight. Lifting it out is necessary in such cases if the child is not to get stuck for a long time; it cannot get loose and be born until the belly bursts. So, as I said before, the woman would have to fear suffering the misfortune of having her rectum injured. Thus a midwife can be of assistance in this circumstance, but must act judiciously.

CHRISTINA: God help us! How many and sundry are the complications and dangers a woman faces in labor. I am nearly losing all desire to become a midwife.

JUSTINA: In our profession our motto must be, Fear God, do right, and fear no one. You know it is a godly profession, and if you practice it carefully and assiduously, as well as is possible and to the extent that God blesses you in it, you can expect reward on earth and in heaven for it. The many difficult births I shall present to you should not keep you from it. I am telling you about them not to frighten you, but to warn and urge you to be careful. And they are not even that common. You may have forty or more happy births before one of these comes along. Otherwise they would never manage, those many inexperienced midwives who when quizzed know nothing but how to cut the navel string. Nevertheless, Our Dear Lord usually fosters a happy outcome. It is thus God's work, though knowledge cannot be spurned, for God's mercy and blessing must abide with well-founded knowledge. As soon as we flatter ourselves that we are something special with our knowledge, we squander God's blessing and become blind and foolish. This often happens to those women who place too much hope in famous and experienced midwives and think they can, with their help, have children without pain or at least have happy births and healthy children with less pain. If it does not turn out that way, then they blame the midwife. Even with all my many efforts and my diligence, this has happened to me. But what can you do with such unreasonable people but be patient and preserve your good conscience? Now when a woman I have never served before asks for me, I am resolved to say—and I make a habit of doing it too—"If you place your confidence in me, then my conscience commands me to serve you. But you must not have a false opinion of me and must not think you will have an easier and happier delivery than you have had and than God wished you to have." It often comes to pass that a woman will have various difficult births, one after another, if her body is not well suited for giving birth, causing her children to get stuck for many reasons. If such things are prevented with well-founded knowledge and God's blessing, the woman naturally gets through it easier than before. In contrast, it can also come to pass that one or another woman has had five, six, indeed even more, happy and easy births, and God nonetheless thereafter sends her a difficult one so she loses her life giving birth, especially if attended by a midwife who does not know how to turn the child. And even when the midwife can turn them, these children, who lie wrong, nevertheless have to die, and the woman has such a difficult birth that she barely survives. Although to this day I have managed to deliver every mother of her child (though some cases were easier than others) and never allowed a single mother to perish with her child inside her if she would but let herself be governed (for which I thank God Almighty), I can nevertheless not guarantee any woman that I can save her if God has ordained death for her or her child. God can make the sighted blind and the blind see. He can just as soon help without means as with means. So I will not trade on my reputation as evil people accuse me of doing; they say I claim I can deliver women without pain and say I guarantee the survival of woman and child. These are nothing but false lies. For even though I can say with good conscience that I have yet to encounter a child that could not be removed from the mother, I certainly will not claim I could never encounter such. It would be a presumption against God. I have on occasion removed dead infants from women, sometimes easily, sometimes with great difficulty, and the mother nevertheless died thereafter. I have also frequently attended easy and happy births, but nevertheless during the lying-in period something came over the women and children and they died. Thus human life is in the hands of the Lord before, during, and after the birth, and no woman can rely on me any more than the extent to which God gives His blessing and mercy.

CHAPTER 3

Of touching in natural births, or when the child is in the right posture. How you know whether a child's head lies right or is fixed on the sharebone or lies too hard against the backside or lies too far to one side or is too large or whether the shoulders are too broad

CHRISTINA: Dear sister, tell me the true cause of one woman having a harder time giving birth than another.

JUSTINA: The true and fundamental cause lies with Our Dear Lord who has everything in His hands, life and death, fortune and misfortune. And His own words proclaim, "In pain thou shalt bring forth children."[70] Perhaps it is a trial placed upon pious Christians that makes it harder for one woman than another. Our Dear Lord alone knows best why He often forces the pious to bear crosses and lets the impious off scot-free. I will tell you that there can also be many natural causes for it; these can be averted with sound knowledge (if not entirely, then to a large extent if God gives His blessing), so the birth will not take so long. The difficulty of passing through cannot be prevented. But, if you have sound knowledge of these things, you can certainly prevent the fetus from getting lodged and fixed as well as prevent the inward mouth from being pulled out[71] from the force of the child's head, which is very critical. The lingering labor and danger that frequently occur will thus not ensue, inasmuch as all this can in fact be said to be part of a natural birth. So pay attention to the differences that need to be noted; you will be amazed, because first of all, when the inward mouth is so firm and rigid that it cannot yield to the opening necessary for a complete birth, it will cause a hard birth, as I said before. Second, it will be a hard birth if the inward mouth lies too close against the rectum and is drawn upwards by the fetus. This is common when the womb is tilted too far forward. Third, it will be a hard birth if the child comes headfirst but goes or threatens to go off to one side or the other. Figure b shows how the child can be helped when the waters break, if it could not be guided when the waters were still intact, as often occurs when the child lies up too high. The next engraving, figure c, shows the head in a bad posture fixed on the other side of the sharebone and with the child lying on its back. Thus an unhappy delivery for mother and child can and must result if the child is not aided immediately when the waters break. This aid can occur manually as you saw in the previous engraving, figure b; the head must be guided by the hand from the other side, as figure 3 [in the second series of engravings beginning on p. 91—*trans.*] indicates, where the midwife grasps the child by the back of the head. A natural birth will ensue, but the child presents with its face upward, which often happens, as the next engraving, figure e, illustrates. This birth is to be sure more difficult than when the child presents face down, as shown in figure d. Nevertheless, it is better for the head to present in some fashion, and it seldom imperils either the mother or the child. Sometimes and sometimes not it comes to pass that, when the head lies crooked, the entire body lies crooked and lurches backwards. Because the child has not yet experienced

70. Gn 3:16.

71. A second-degree prolapse, that is, when the cervix protrudes beyond the introitus.

the force of the throes and the waters have not yet broken, it usually lies with only its head crooked, as in the here-cited figure 3, where that sort of posture can be seen. If the child is not aided soon after the waters break, as described above, a posture like the one in the previously cited figure c will result, because when the waters run off, the force of the pains and the curling up of the child gradually cause it to come down on its back. Because its head lies to one side and cannot enter the birth passage, as figure c demonstrates, nothing can be done but turn the child by its feet and save the mother, since the waters are too far gone. If a child lies right, as shown in figure d, then it will exit felicitously of its own accord, driven out by Nature. Children may also lie in the right posture for birth, as figure e indicates, but they enter the birth passage face up; soon after the delivery begins, the child lies on its back. When such a child lies face upward with its head crooked so it cannot enter the birth passage, the same kind of danger as that depicted in the oft-mentioned figure c ensues, unless its head be helped into the right posture when the waters break. And even if you assist it immediately, it is still harder to deliver it than an infant in the right posture—as described previously—because the child is lying on its back with its face upward. But it is not dangerous for mother and child, and you cannot do anything about it except help the head into the birth passage right away.

Children can also be in the right posture for birth, but their shoulders are too broad. Figure f illustrates this and how the midwife can assist: she puts the fingers of both her hands on both sides of the child's shoulders in order to squeeze them together as much as possible so it can get through far enough for you to reach the armpits and take hold of the child under its arms. You can pull on it then, aiding the throes by pulling so the mother is saved nevertheless. For if you let it run up against the sharebone with both shoulders, the throes will push the child down into a ball. Peasant midwives say then, "The bear-mother has locked up the child. It cannot be born." Generally people say, "The woman's belly closed up." In figure f you can easily discern the closure and you can note how such a large child, if it is not carefully attended when the head pushes through, can get stuck. Of this unfortunate condition it can rather be said that the belly was stopped up rather than that it closed up.

A child can present properly with its head, and yet when the head actually pushes through, the child's entire body lies crooked and its shoulder fixes on the sharebone, as in figure g, which demonstrates manual intervention.

The following engraving, figure h, depicts a child that is certainly lying right, but its head turns the wrong way during delivery so it presents with its face. It shows the midwife's hand, as you see. She helps the head up be-

Figure B

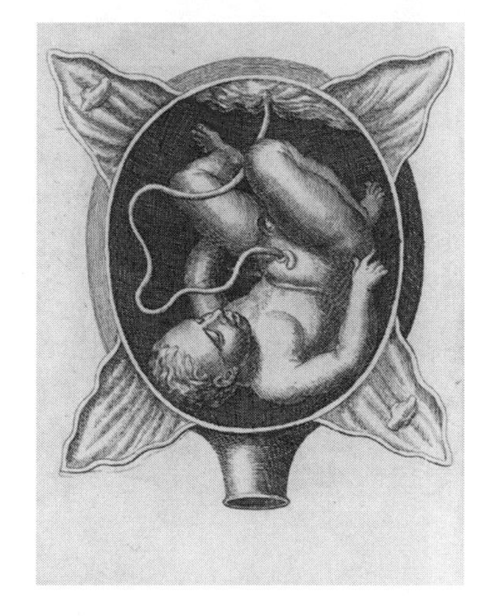

Figure C

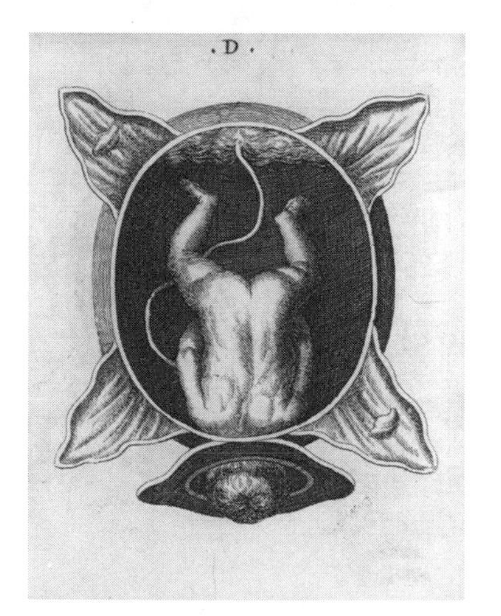

Figure D

Figure E

cause there is no other way to help it. Various times, when the waters were still intact, I have tried to push back the infant's face right away so I could pull down the head properly as children in the right posture are wont to have their heads positioned. But it does not work. Children in this posture have very little water round them and generally have been pushed into this unnatural posture by some kind of violence that has befallen the pregnant woman. I have variously ascertained that pregnant women suffer such com-

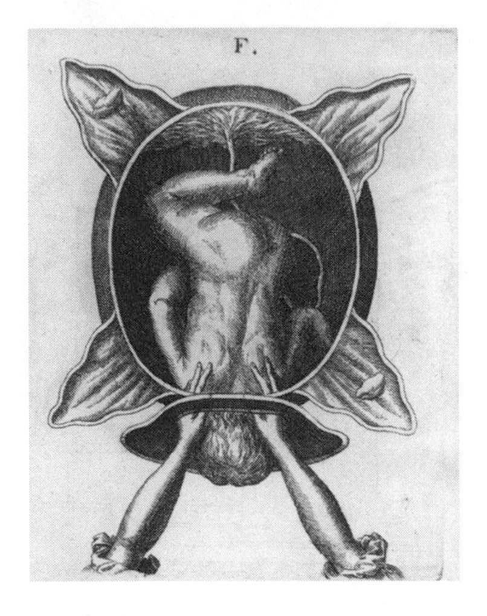

Figure F

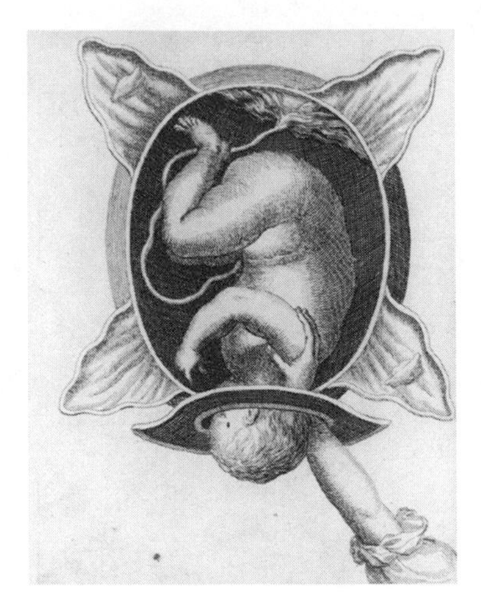

Figure G

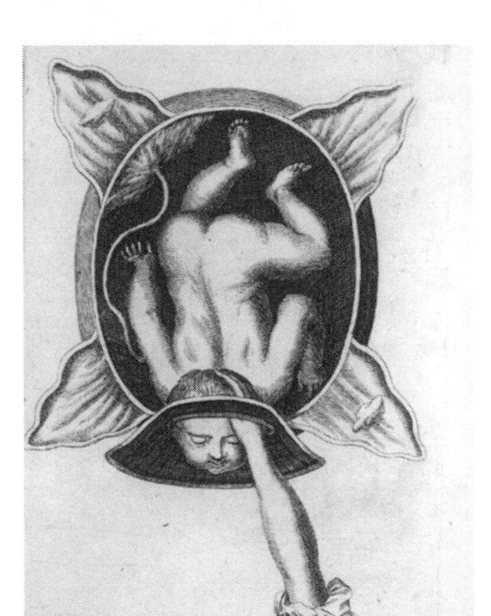

Figure H

Figure I

plications as a result of a violent jolt of the horses and carriage. I myself have ridden along with them. Before we get in the carriage, I have for certain reasons checked to see whether the children lay right for birth and whether they had already turned. And I have found them lying exactly right for birth before the ride. But after the ride and the violent jolting, I have found them presenting with their face. Because the women complained loudly about the violent jolting, I needed to examine them and therefore was allowed to touch them. Children also have remained in such a posture, the one for three weeks, the other for two weeks, until the hour of birth arrived. It also frequently comes to pass, as in the various cases I have already pointed out to you, that though children are in the right posture, they are nevertheless quite imperiled and in danger of losing their lives. I want to show you an infant lying right that nevertheless is quite endangered: the cord or the navel string is coming out in front of the head. Take a look at figure i and its explanation, which can be found at the ☜ sign in chapter 7, where the necessity of breaking the waters is outlined.[72]

Fourth, the birth is still more difficult and dangerous if the infant lies wrong for birth, as you already saw in figure c and as can occur in various additional ways, and as you will come to know if you pay attention.

CHRISTINA: How can I pay attention to this? It is hidden, and I cannot see it.

JUSTINA: Of course you cannot see it, but you can discover it when you touch the woman.

CHRISTINA: How can I feel it when I touch her? Some women labor for a whole day, and I cannot get anywhere.

JUSTINA: You have to know that the mouth of the womb lies deep inside. You cannot get anywhere because you are not yet familiar with it. I shall describe it to you in detail, however, so you can avert danger before it arises, which is possible to do. If a woman endures a day of pains, then something is certainly causing it that you need to know about and prevent. At least it is necessary to know whether it is true labor or not, as well as whether the infant's head has gotten fixed somewhere. So the issue is, how to recognize true and false labor pains, as well as how to tell the difference between a right and a wrong birth. The mouth of the womb and touching the infant through it will give you everything you need to know. If you do not pay attention, then you are stuck; the longer the pains squeeze a child who is fixed, the less likely you are to get it free thereafter. As long as it is still alive and has some energy, an infant lodged like that will stay high up in the womb and draw the opening up toward itself. If the child is small and strong enough to force the mouth open wide enough with its own motion, a violent pang will force the opening, as well as push out the child. And, so you think, everything is taken care of at once and the child born. But you are deceived in this thought, for you are not yet familiar with the inward mouth of

72. The hand sign referenced here does not in fact appear in chapter 7, nor is figure i mentioned there, although, as Siegemund promises here, the chapter most assuredly outlines the necessity of rupturing the membranes should the umbilical cord present.

the womb. If a large infant gets lodged, then it is touch and go whether, being still alive, it can get free on its own. If, however, the pains have at it, it will go into whatever posture they force it into; they will either flatten it or push it crooked, depending on how much room there is, and so it will fill up the birth passage and yet cannot come forth, and the woman will be most greatly imperiled by it. I have been fetched several times when such danger had arisen; the women had had to labor for two to three days on account of the child's head being fixed. I have found the child dead as the result of it and the mother in the gravest danger of losing her life, inasmuch as the child's head had fixed firmly on the sharebone as a result of the violent pains and of her belly being stroked too hard. The attending women do this; they stroke the belly too early on, intending thereby to help, and the midwives are no better informed. They feel the child's head and yet do not know whether it is fixed more on one side than on the other or whether it has pushed up too much against the rectum or whether it has pressed too firmly against the sharebone or whether the child's head is too large. If a midwife cannot distinguish among these various conditions, she says, "The child lies right and lacks only the throes." But she is much mistaken. The throes in such complications are generally called small and false labor pains, when they are neither small nor false, but rather cannot make use of their force as they should on account of the child's being fixed. When the fetus or infant gets fixed right at the start, violent aid is supplied by means of medicaments, and the woman in travail is urged by the midwife, as well as the attending women, to push. Then the child's head, that is the cranium, flexes, and a rim or ring forms round it (as they call it).[73] Thereafter they sit there until the child is dead. Indeed, the woman probably will lose her life over this, especially if it is a large infant. If the midwife does not know what is wrong, she will also not know how to help; otherwise she would have averted the danger in the first place. This will serve to inform you: you can perceive the state of things by means of a thoughtful manual examination.

CHRISTINA: Dear sister, acquaint me better with what you have told me up to now.

JUSTINA: I shall do so at length, insofar as I am acquainted with it myself, for your benefit and that of my neighbor. If you think a woman may of-

73. Siegemund most likely refers here to pressure leading to a swelling resembling a chignon at the back of the cranium, known as *caput succedaneum*. Note that she focuses on the rim where the swelling begins rather than the swelling itself. It is also possible that she simply means bruising or swelling around the head resulting from pushing against the pelvic inlet in labor with a contracted pelvic inlet. I thank Amy Ravin, M.D., for explaining *caput succedaneum* to me and for showing me live examples.

ten labor for a day and yet you will not be able to feel anything, you are mistaken. If a woman should labor for a day (I say labor—not vain or false labor as it is called—as often comes to pass with women laboring extremely hard for two days), there has to be a reason for it. I have been fetched all too often and had to separate midwives and women in travail,[74] and the women then go on for four or five more weeks. What kind of knowledge is that then for a midwife, or rather what is wrong with her that she does not know what the mouth of the womb is; otherwise, she could easily distinguish between true and false labor. I want to talk about birth in general and show you your error concerning the opening of the mouth since you believe the child would come all of sudden with a violent pain, though labor has gone on for an entire day. Just learn how to examine skillfully the inward mouth of a woman in travail right at the beginning; this mouth will teach you otherwise. You will find in all women an opening at least fourteen days, even four and more weeks, ahead of time; if a woman is disposed to give birth easily, all the greater and earlier the opening. However, because with many women the inward mouth lies deep inside and toward the rectum, you can easily be mistaken. For it opens so high up that of course when the opening is large enough and cannot hold the infant back any longer, the infant will suddenly enter the birth passage with a violent pain. But if the child is in the wrong posture, then God be merciful, for nothing can be done other than save the mother and get the child out with as skillful a manipulation as is possible. If it is lying in the right posture and has not fixed anywhere, then it is a very good thing, and it will emerge upon a violent pang. I advise you not to be too trusting. I am for touching the mouth as I instructed you in chapter 1. Thus you will find the mouth cannot open, as you think, after one or two pains. There are to be sure women who have children after two or three pains: such people have during their pregnancy six weeks or more before their delivery a complete opening. So I often say if the woman should cough violently, her child would fall out. But there are not many bellies like that, so it is easy for a midwife to know before a woman goes into labor whether it will go quickly or with difficulty. Thus I think it best for those I have anything to do with to summon me in a timely fashion so I can find out how things stand with mother and child, especially in the case of bellies prone to difficult births. Necessity will no doubt teach such people to seek advice in a timely fashion, and no one will have to ask them to do it; they will no doubt come on their own. Those women who have tolerable labor have no need of these precautions. But it is better to be good and cautious than to be

74. That is, tell the midwives to leave.

overconfident. Sometimes a woman has had three or more happy deliveries and then has a hard birth come upon her. I advise you once again to touch the mouth in plenty of time so you will know for sure whether the fetus is lying right.

CHRISTINA: What should I understand by "touch the mouth in plenty of time"? Some women are difficult and reluctant to allow themselves to be touched until they are in the most dire straits.

JUSTINA: It is not necessary to torture such women ahead of time, and no woman will allow it until she is in labor. But when she is in labor and there is pain, you must not wait too long, be it true or false pains. You will be able to distinguish them by touching her and will be able to say what is to be done. If, however, it happens to a peevish woman who, stricken with fear, will not allow herself to be touched, then be it upon her own head come what may, and she cannot scream afterwards about the midwife, should something bad happen. He who will not take advice cannot be helped. A woman cannot easily go into labor without the mouth of the womb being slightly open (as noted in chapter 1). This opening is even at the commencement of labor wide enough for two fingers to enter, even with women who are experiencing the most difficult labor, unless it be a woman giving birth for the first time, as mentioned above, or a midwife with clumsy fingers, which is not good either.

CHRISTINA: When I find an opening, how do I determine whether the child is lying with its head in the right posture for birth or if the head has lodged somewhere?

JUSTINA: When you have located the womb and its opening, touch it with two fingers as gently as possible and feel inside the open womb. If the infant lies right, you will soon come across the head. You recognize the head in no other way but that it is round and feels hard, and when the child moves, you will feel the opening in its little head;[75] you must not put your finger in it. But if you want to have a deeper knowledge of it, you can without doing any harm move your two fingers like a compass around the child's head; it will not hurt the child if you but do it gently. The head can be felt, but you need to be careful of the skin that encloses the infant and the waters, and that is part of the afterbirth, especially when the woman has labor pains, for then the skin expands and becomes so taut that if you touch it roughly during a labor pain, the waters will break, though one skin is tougher than the next. In time you will come to know how to touch it. But I prefer for you to do it too gently rather than too roughly. If such throes

75. Fontanel.

come when you are with the woman, you must not withdraw your fingers, but you should not press on the waters; rather keep your fingers in the mouth and before the waters. You will thereby soon get to know true labor pains. As soon as the throes come, the waters become hard and violently push through the mouth. When, however, the pains subside, the skin becomes soft like a little cloth, and you can actually feel the head. When the pains are there, you cannot feel the head so well because you cannot press on account of the hard waters for fear they will break too soon, for if the waters break too early on, it causes a difficult birth. Thus you must pay all the more attention when there are no pains and feel with your two fingers around the entire head whether it is descending straight into the birth passage. You will soon notice whether it is lodged to one side more than to the other. When there are no pains, it can be moved as you will. You should guide it in straight so when the throes come they will seize it and push it straight into the birth passage. Do that until you find it is coming forth straight. It also sometimes comes to pass that while the head is being guided in, the child tries to retract it. When that happens, it is to be feared that the child will turn if it has enough room. So you need to leave off guiding the head until the waters break. If they do not break on their own, then it is necessary to break them to prevent misfortune. If the waters should break in the process and the head is not yet guided in, you can help it all the better. For if the waters are gone, you need not be afraid of harming the child, except you have to watch out for the opening in the head to keep from injuring it, and because there are no throes, you can insert your fingers in there where the head has gotten most jammed so when the throes come, the head will be forced to slip off [from where it is stuck—*trans.*]. You have to keep your fingers in there until the throes have passed unless you feel it will let itself be easily diverted. In that case you must not apply too much pressure; otherwise the child will push too much up against the other side. If you wish to get a proper understanding of the matter, you need to pay attention to and reflect deliberately upon your hand and its feeling. It can no doubt be rightly felt, but it requires precise reflection. This diversion averts the danger recounted above of the child's head fixing somewhere, be it on the backside or on the sharebone or to one side.

CHRISTINA: When I find the sort of complication you have told me about, is it beneficial for me to strengthen the pains and to urge the woman to bear down?

JUSTINA: You need to be careful not to urge the woman to push too hard until you have guided in the child's head properly and it is fully in the process of descending so it can come forth. So beware of strengthening the pains prematurely.[76] For if the head is not guided in right, the pains and the untimely assistance will lead to the other grave dangers reported above, whereas Nature would not push it so hard. So it is also better to relieve it off [of wherever it is fixed—*trans.*], for the sooner the head is diverted, the better it is for mother and child. I will show you, furthermore, the dangers that ignorant midwives and foolish strengthening of the pains can lead to. Read the eighth court testimony, which Frau Barbara Vogt gave, that follows the chapter on the necessity of breaking the waters. She was foolishly forced to labor so hard that the birthing chair broke beneath her. This foolish stimulation of the pains happened on Saturday, the first day, and she was not rescued and delivered until Tuesday along toward morning, and to be sure under gravely dangerous circumstances, as you will find in the testimony, for I did not arrive until Monday along toward evening.

CHRISTINA: My dear, do you think this unhappy birth was caused by the pains being stimulated too early or too violently and that is how the child shifted? The infant may have shifted when the chair broke.

JUSTINA: Seeing as how the woman broke the birthing chair beneath her, as she herself testified, the immediate cause was certainly the result of the pains being stimulated so she was forced to push. Of course this fall itself could have contributed substantially to her imperilment. But to consider the matter properly, the child must still have been lying high up in the womb and crooked with its head up against the rectum, since its head could be shoved and displaced so violently as a result of the fall or from pushing too hard, which predicament would have been impossible if the head had been positioned properly beneath the child right in the inward mouth toward the sheath. Be careful of this. For if the child's head has not yet really penetrated the inward mouth and neck, no midwife can stimulate the pains in good conscience, because when the pains are stimulated early on like that when the child does not yet lie squarely and securely in the right posture to enter the birth passage, the mighty force of the untimely pushing can very easily cause the child to shift so it is flexed and crooked or even cause it to flip over, especially if the waters are intact and it has room to do it. This is what happened to this woman, and indeed it was the result of the ignorance of the midwife. For this woman should not yet have sat on the birthing chair because the infant was no doubt still high up inside her or indeed already lying crooked with its head against the rectum, as the dangerous outcome indicated. Still less should the midwife have stimulated the pains and urged the woman to push. It would have been better to have laid her in a warm bed

76. That is, with medicines, as was customary in many circles. Siegemund is here as elsewhere chary of medicines.

and told her to be patient, to have sat down next to her and gently touched her to see what the pains that come from Nature had accomplished and whether the child's head would permit itself to be guided into the birth passage with the natural throes or not. For if the head lies straight, the natural throes sharply force it into the birth passage so you can really feel it. If the throes cannot force it into the birth passage, then you must gently search for where it is stuck or to which side the head is going off crooked or threatens to. Thus, as I have already instructed you and shown you sufficiently in detail, you will not have to be afraid of caring for women experiencing these kinds of unfortunate complications.

CHRISTINA: What, however, if the child should have a very large head? Should I not push there either or give her something to make her push harder?

JUSTINA: It is just the same if the head is too large. If you push hard and strongly urge the mother to help, then the head will be pressed wider and the danger will become even greater. For the cranium or skull can be molded in the womb, as can be seen from the pointy heads of newborns. It can also be molded so it is wider if you push too early on. It is much better to let Nature take its course; you simply see to it that the head does not get fixed to one side. Thus the head has to become pointy, as is necessary, and there will be no danger except for a somewhat protracted labor. However, if it is extremely slow, it can sometimes cost mother and child their lives, at the very least the child its life and the mother her health. I have often been summoned when women have been laboring for three or four days in gravest danger because the head was pressed so wide that it was impossible for the women to be delivered of their child. And the head, which was stuck on all sides on account of its size and because the pains had been pushed too hard, could not get through, and the child had to die of it. Indeed, I found the child already dead inside them, as indicated by the stench and its peeling skin, as well as by the completely collapsed skull, which had been so crushed in the birth passage that it was a sight to behold. However, because these women's strength was completely spent because of the time it took, it could rightly be said that the child presented for birth and there was no strength left to bear it. Because the stench of the dead child saps her strength excessively, it is difficult to rescue a woman under such conditions. Certainly God helped me save all those I treated in this and other ways, which I will further cite. Nevertheless, the sooner I got there, the easier it was to save them. When, however, I was slow to arrive, I knew of no better way of saving the woman, the child being dead and rotten and she having spent all her strength so she could no longer help herself, than of fastening a hook in the child's head in the split or crushed skull. Inasmuch as the head was completely split in two, it could be pulled long ways.[77] Nature thereby gets a second wind—but a wind with a terrible stench—and begins pushing, even if the woman herself can no longer move a muscle. The pains are also no longer necessary, for the belly is so pulled apart from the preceding violence and the throes that the infant can emerge with pulling and light pressing, as long as its head is turned long ways.

CHRISTINA: What, however, if the skull broke to pieces?

JUSTINA: I have had the skull break loose on me; you have to be careful with complications like that. For when you make use of the crotchet, you need to make sure you stick it in so it cannot be torn out and so it does not poke out the other side. Otherwise you will injure the woman. Likewise you have to be careful when pulling; if the cranium breaks to pieces or the crotchet rebounds, the woman will be injured by it. I speak here of the pointy, clumsy hooks I made use of at first. I knew of nothing better in the villages among the poor peasant women; I imagine there are plenty of midwives who cannot and do not always have suitable instruments with them, either as a result of their own ignorance or because they simply lack them. Keeping such circumstances in mind, I shall describe to you at length how to use the crotchet carefully and shall indicate in the engraving [in chapter 9—*trans.*] with the birthing bed the pointed as well as the broad crotchets that are easiest to use. Once when the cranium broke on me, I had my right hand or rather fingers at the child's head where the crotchet had been stuck in. Paying careful attention to it, I felt the cranium breaking and left off pulling on the crotchet with my left hand, and completely withdrew it, along with the broken cranium. You could easily break off one piece after another, insofar as the head was completely rotten. And God felicitously aided me so when the cranium was out, I could use my hand to pull out the rest, since it was now reduced in size, and to this day the woman lives hale and hardy, still bearing children. In such a circumstance I cannot guarantee the life of any woman. For when Nature is so weakened that the woman cannot recover, she will die afterwards whether or not she is freed of the child. Some of them have died on me a few days later. One person can endure more than another, especially if God wishes to preserve them. So I hope I have told you enough about the head getting stuck as well as about children with heads that are too large—they are treated in one and the same manner—so you can see what danger can result from hard pushing. Practice will

77. Siegemund probably means to indicate by "long ways" that because the skull is crushed and no longer as long front to back as it would be normally, it is no longer necessary for it to rotate ninety degrees as it would normally during passage through the pelvis. It can thus be extracted in the transverse position (with respect to the pelvic inlet).

teach you how to guide the infant into the birth passage, and you will have reason to thank me.

CHRISTINA: Do tell me how I can learn to prevent the danger you have told me about and what I have to watch for with it.

JUSTINA: If you are taking all manner of precaution with the child's head, then you will never remove your two fingers from the inward mouth. I am not saying you should attend the woman in this manner too early on or for too long a time; you [just have to do it—*trans.*] in plenty of time once you know how the child lies. If there are no complications, then you must not do anything. If there is a complication, as I reported above, you must do your part when the labor pains nevertheless continue—not that you have to put the woman on the rack or never leave her side. Instead, when you come to her and stay resolutely with her for five or six pains, then you can leave her in peace again for a few pains and resume helping her thereafter so she does not weary of you and also so you do not make a mistake while you are at it. When the throes come that lead to the child's passing through, then it is advisable for you not to get up. Time will certainly teach you that. You can leave the woman lying in her own good bed until the throes come that lead to breaking through, even up to the last one. If you sit down with her a few times in a good bed, the guiding of the infant can take place there just as well as in a birthing bed. You can also have her get up if necessary—especially when it is still early on—to try and see in what position the throes are the most effective, and you can be guided by that. Sometimes lying down inhibits the pains; standing often does as well. A test cannot hurt, especially if it is done early on. A fetus that is fixed somewhere can also more easily be guided with the help of a labor pain into the birth passage when the woman is standing than when she lies down a lot. Experience will teach you. Try it in all manner of ways. But a woman cannot endure standing for long. If at the longest she cannot endure it for two or three pains, then have her lie down. It is not as dangerous should the woman faint. If you undertake it with caution, it will proceed without injury. Enough then of complications that midwives mostly take to be natural births.

CHAPTER 4

Of the wrong postures of children, how to recognize them and how to assist or manage each of them, including turning the children

CHRISTINA: If what you have told me about up to now are called natural births, I would really like to hear how unnatural births differ and how to assist with them.

JUSTINA: I will outline the knowledge of them I have acquired from working with women who summoned me when there was nothing left to do but save the woman. I will also tell you about those I attended starting with the onset of labor and stood by until it was over. I cannot possibly manage to bring up all of the people I have treated and therefore feel the need only to present the postures of odd births, insofar as they occur to me and insofar as I can recall those that have come under my care and how I managed them.

First, I will talk about saving women to whose side I was summoned when their children were already dead, and whereas there are many variations of the arm presentation,[78] I will describe them as I found them and tell you how I managed some of them easily and others with great difficulty. I will start with my first case.

(Of a child that presented with its little arm or hand, see figure 1 [in the second series of engravings beginning on p. 91—*trans.*) The first child I helped deliver was one that lay for fourteen hours presenting with its arm and little hand. Half its arm with its little hand was hanging out of the belly. The woman was in her third day of labor, but the child's hand had not shown itself right away. Since the situation appeared to be very dangerous, the midwife doubted the mother and child could be saved. The midwife was the sister-in-law of the woman in labor. She placed all her faith in me, inasmuch as in her eyes all human aid was at an end. The woman had six living children to rear, which heightened their despair. They were poor peasants. The midwife, that is, her sister-in-law, entreated me, for the love of God, to advise them because she had seen me with books with illustrations of sundry births. So I got out the books and looked to see what postures were depicted there. Because, however, it was impossible for this midwife to determine which picture corresponded to the posture of the laboring woman's child, they despaired. The sister-in-law, however, had great faith that God would save her child through me because she knew I had pretty well worked my way through the books. But I was still young—and I have to say quite foolish—to make that kind of attempt, for I was but twenty-three years old and had not yet had a child myself, except that two years before they had said I was pregnant and for forty weeks I had gone round with that idea. All the mid-

78. Siegemund uses the word "hand" loosely, sometimes meaning the hand and the arm and sometimes merely the hand. When referring generally to the problem of the emergence of the hand or arm before the head, I have chosen to use the term "arm presentation," even though Siegemund actually uses the German word *Hand,* as the former term is more current in English and also comes closer to what Siegemund actually means. When I judge Siegemund to refer more specifically to the hand, as opposed to the hand and the arm, I use the English word "hand."

Figure 1

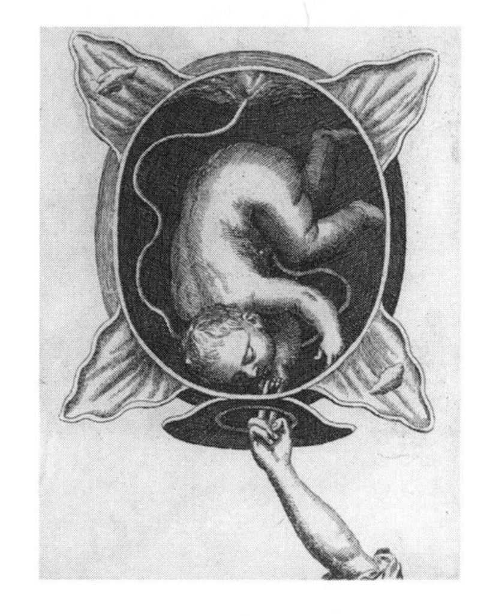

Figure 2

Figure 3

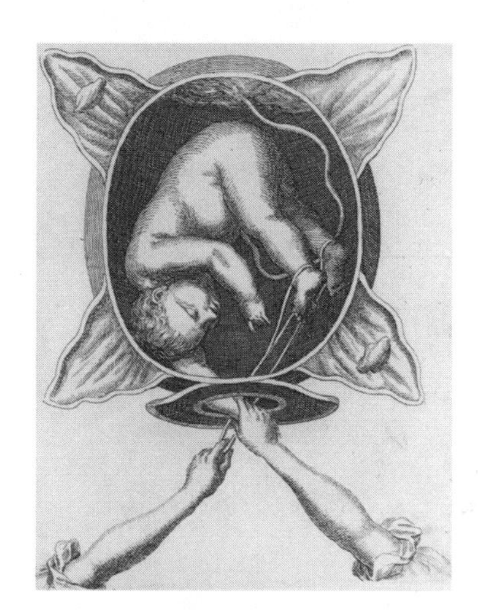

Figure 4

wives thought I was pregnant. My time finally came, and the midwife was fetched who said the child lay right. So I labored in this manner until the third day, but no child came forth. We did not have confidence in this midwife, and another was fetched who confirmed the child lay right. She kept at it with me until a third and thereafter a fourth midwife were summoned who agreed with one voice that I was to give birth and the child lay right. Thus I was tortured for fourteen days and was kept on the rack in all man-

Figure 5

Figure 6

Figure 7

Figure 8

ner of ways until I was at death's door. Thereafter I was left for dead when Our Dear Lord took pity on me. The midwives still insisted I would die along with the child. Since no one was looking after me, God preserved me and unexpectedly sent the wife of a soldier into the village where I lay. She was also a midwife. Since people were lamenting my situation, especially on account of the great misery I had borne only to have it be all in vain, this soldier's wife demanded to know the condition I was in. Thereupon my hus-

Figure 9

Figure 10

Figure 11

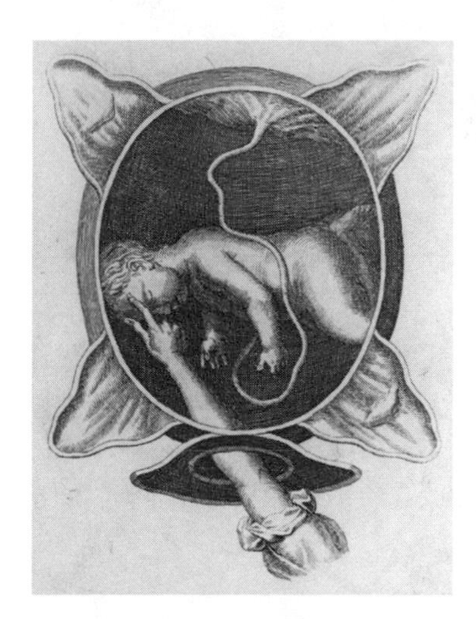

Figure 12

band and my mother summoned her and asked her to state her opinion. When she had gained a good understanding of what was happening with me, she said I had no child and was by no means pregnant. It was a clogging of the blood and a serious disease of the womb and a fallen womb. Given this opinion, a medical doctor was accordingly summoned for advice. He cured me with God's blessing. This dangerous experience provided a great inducement for me to spend time thereafter assiduously reading those books. The

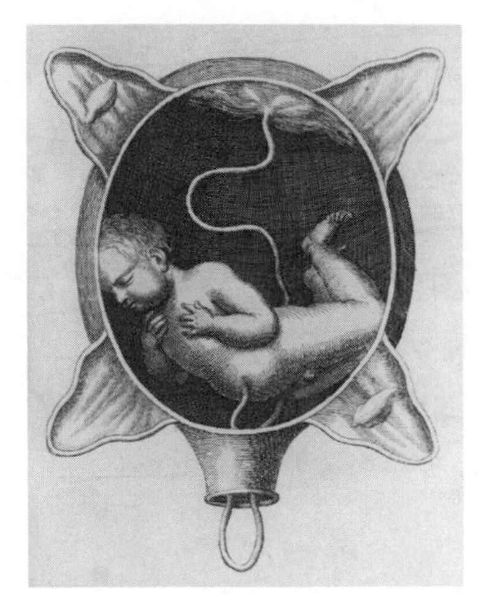

Figure 13

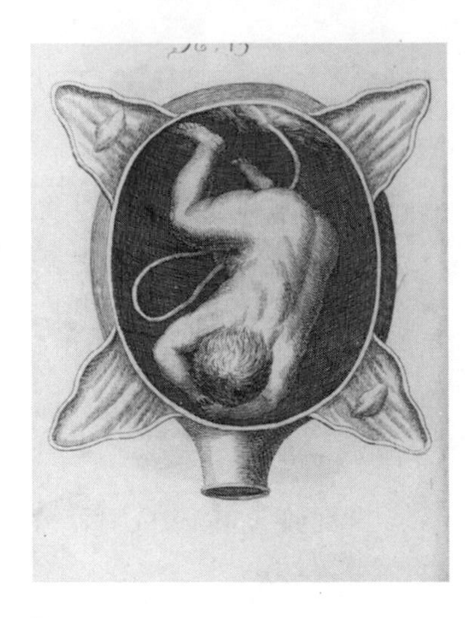

Figure 15

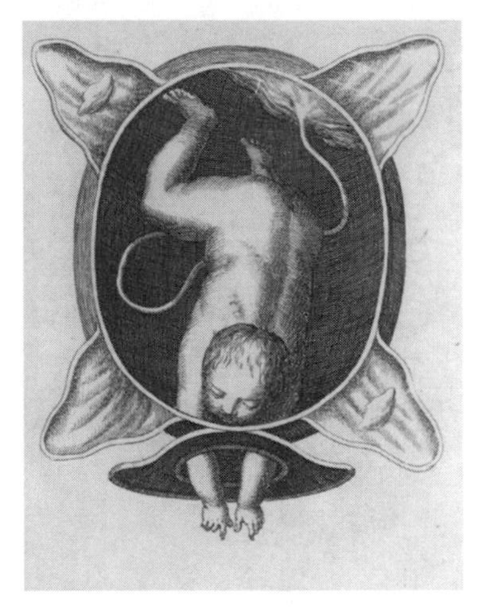

Figure 16

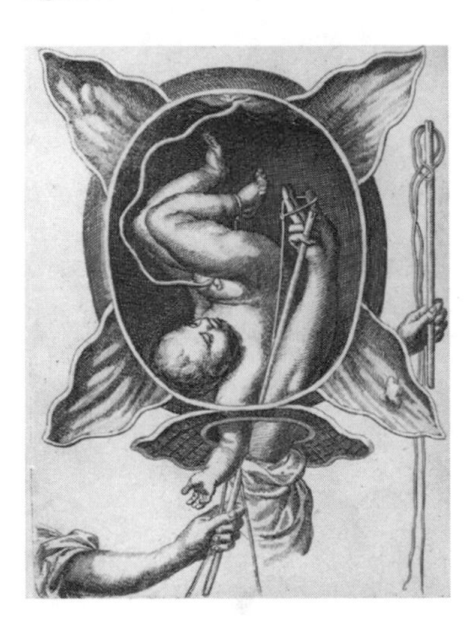

Figure 17

midwife, the sister-in-law of said woman in labor, knew all this and had herself been present during my fruitless labor. She expected good things from me as a result of the books I had read, and I too had a great desire to learn something out of the ordinary, and they were peasants who did not know how to seek additional assistance and yet were of the opinion that the laboring woman would die before their very eyes. So I let myself be persuaded to touch her to see whether the little hand could be brought back inside. Us-

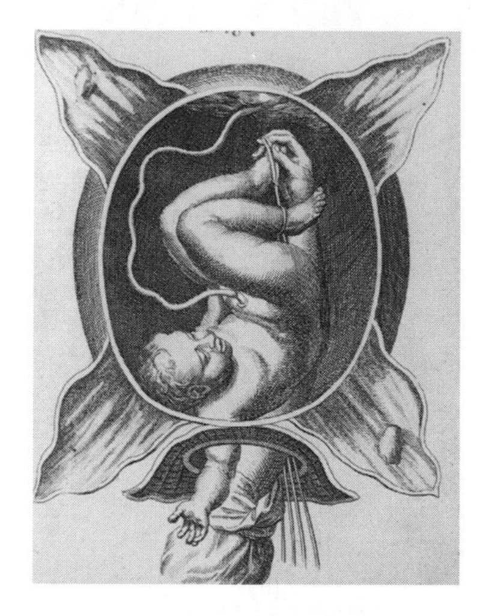

Figure 18

Figure 19

Figure 20

Figure 21

ing warm beer and butter to lubricate it, I very gently and deftly put it back in as far as I could (see figure 1, showing how I put the child's bent arm back inside and pushed its arm back in by the elbow.) Its little hand could be gotten back deep inside, but I did not know whether this was good or bad or whether the child against hope still lived. It then pulled its little arm in toward itself, and as a result of this movement its head moved into position before the birth passage. Thus a natural birth ensued, but the child was weak.

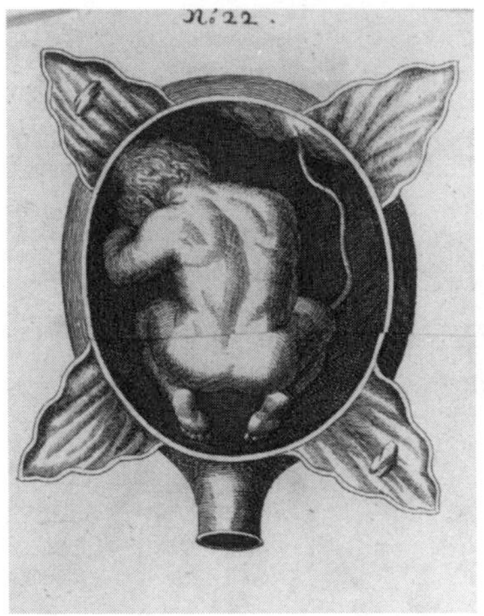

Figure 22

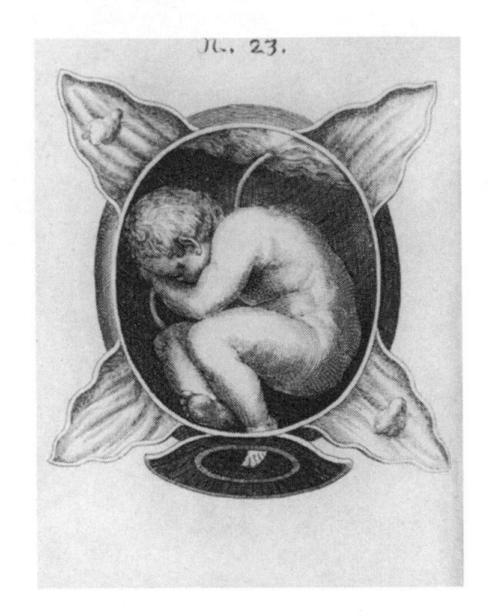

Figure 23

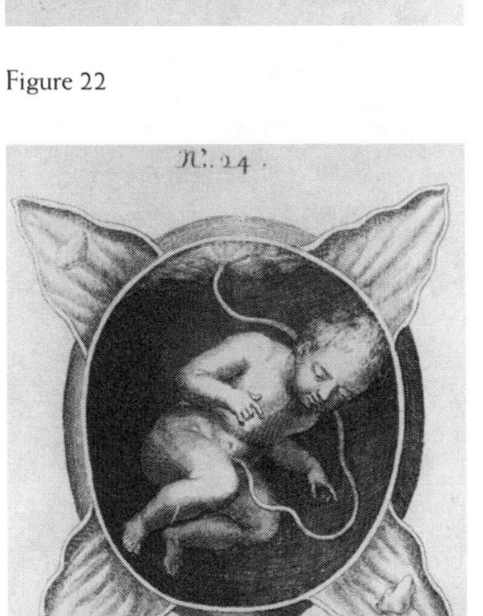

Figure 24

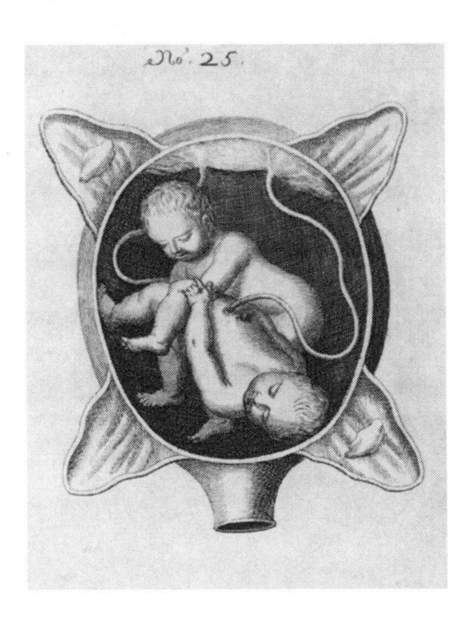

Figure 25

This birth went well for me, though I did not know what I was doing and I was none the wiser for it. Now, because I have more knowledge of these things, I think the child's head had caught on the sharebone and was crooked, so nothing entered the birth passage. When, however, the child was able to retract its little arm once I pushed it back inside, its head was forced to shift too. For it is said: if it is God's will, even a cock can lay an egg.[79] This birth did not make me any the wiser, except that I learned that the child's head fills the entire birth passage, that it is as hard as it is, and that it can shift closer to the opening. With that, I was overcome with a desire to learn more, and the midwife faithfully helped me do so, summoning me to the side of all the women giving birth she attended, and I was surprised how well it went for me. And I became proof of the saying: burdens become light when cheerfully borne. And so I gained a reputation among peasant women having difficult births, and they fetched me to other villages. I must therefore confess that I gained most of my knowledge through practice with all kinds of unnatural births. Many people summoned me to assist for the sake of God, so I made every effort to extract the child and save the woman, inasmuch as in the cases where I was called in, the children were usually already dead. And insofar as mortal aid was possible, things always ended happily for me with God's blessing, may God on High be thanked. May God continue to aid me!

To continue to detail arm presentations, I now have greater knowledge of what can happen if the hands come forth first and how they come forth in such unhappy births, as can be seen in the first birth I have explained, where it without a doubt started out like that. If the little hand had been put back inside when the waters broke, the woman would not have had to suffer for so many days.

But to explain to you how arm presentations can be prevented when the birth is close at hand and the waters are unbroken (as figure 2 demonstrates), you must pinch or squeeze the child's fingers with your fingers. It will then pull back its little hand on its own. If this happens with the waters still unbroken, the head will often shift on its own into the proper place for birth when the child pulls back its little hand, which thing I have tried various times. If it does not do so on its own, you have to help it (figure 3 portrays the intervention.) You have to intervene by pinching its little hand when there are no pains, otherwise you will break the waters. When there are no pains, the skin that encloses the water is quite soft and can be pressed together like a rag over the child's little hand or fingers. You must not pinch the child's fingers with your fingernails, but squeeze them with your fingers. I call it pinching, as you can see in figure 2. If you do not do this immedi-

79. The German original saying is closer to the critical situation of birth: "Was Gott will erquicken/ kan sich nicht erdrücken" (What God wishes to revive, cannot be crushed). In other words Siegemund refers here to the ubiquitous power of God as in Rm 8:31: "If God is for us, who is against us?"

ately at the onset of labor or at the point when the waters break, the child's little hand will penetrate farther into the birth passage and its head will be crooked to one side, as you can see in figure 4 where the waters have already broken. It is harder to intervene at this point than to prevent this from happening in the first place. The sooner you help—before the child is squeezed even harder from the pains and must die—the better it is for mother and child. I have turned many living children, and they have survived. They were, to be sure, born feet first, but happy births ensued. They survive, however, only when women are not disposed by nature to difficult deliveries, for otherwise children seldom survive. But it is in God's hands to preserve what He will. In such cases where the child lies wrong, it is best to turn it immediately as soon as the midwife can, for you cannot hope or wait for anything more if the waters have already broken. So I wanted to show you this timely prevention and opportune turn in these four copper engravings, that is, 2, 3, 4, and 5. If the children are still living, you need to set about turning them in this manner. If the child's feet do not lie close together for pulling them out, then you have to put a loop around the first foot you find (as you can see in figure 4) and take hold of the cord or ribbon with your left hand so you can keep hold of the little foot you have gotten hold of so it does not get displaced while you search for the other one. When you have gotten hold of the other one, put a loop around it too. Then you can assist according to the intervention in figure 4 where you take the child's arm or hand with your right hand, bring it back inside and thereby help the child back deeper inside. Then take the cords with your left hand and gradually pull the feet down below the body. In this manner you will easily manage to get them round so you can get the feet into the birth passage as is illustrated in figure 5. If the child is still living (which is often the case), remove the cords with the loops from its feet.

CHRISTINA: If you must put loops around the feet of a living child in order to do it, why do you not pull its feet down with your hand into the birth passage? It would be easier than putting loops around them in the belly. Pulling on the ribbons like that when they are around the child's feet does violence to the child if it is still alive.

JUSTINA: Greater violence would be done to the child if I were to pull it with my hand. I put loops around the feet to keep them together and turn them. If I should pull down its feet with my hand, as you indicate, I would do the child greater violence than when I put loops over it and gently pull on it by the ribbons. If the child is still living, it does not lie so tightly compressed as when it is dead. You must not pull so hard on it as to do violence to it as you would with dead children whom it cannot hurt. The woman can thereby be aided. Thus both things are justified. Because my left hand can do this gentle pulling outside the belly, as you see in figure 4, my right hand can all the more easily put the child's arm back inside and better help the child back deeper into the belly, as you can see in figure 5. The child cannot become as compressed as when I use my hand to pull the feet down below, for I cannot get both hands in any belly. How then should I pull the child by the feet with one hand and use my arm to put the child's arm back inside? As you see here, it can be done with the ribbons since there is room enough for it. When you have gotten the child's feet out, as figure 5 demonstrates, undo the ribbons, as I said before, if the child is still living, and pull on the feet some when the pains come; pull on them as hard as the pains push, but by no means harder. Thus you will not hurry either mother or child too much. You can well aid Nature when you are careful, but you can also hurry it too much and pull too much. If you do not yet have its little hand back up good and high above the sharebone, then hold its feet with your left hand again so you can introduce your right hand deep inside and push the child's hand as deep back inside as you can. Thus its feet will come forth better when the throes come. Do not pull on its feet without the labor pains, for it does no good and does more injury to the child. Make certain too that you guide the feet and the child, if possible, so the child's belly comes forth facing the woman's back. Thus the birth occurs in a manner better for mother and child than if the child's belly faces the mother's belly. For when it gets out up to its head, the child can fix on the sharebone, as you can see in figure 21.

CHRISTINA: You say when I get out the child's feet with the cords and the child is still living (which is often the case), I should remove the loops from its feet and carefully pull on the child's little feet with my hands? I should also be mindful that the child's feet could be injured by the pulling if they have loops around them and so you are telling me to pull gently with my hand? I can well imagine then that there is a lot to be said about pulling in the case of children who are still living.

JUSTINA: You need to know here that I put loops around the feet only when I am forced to as, for example, when I arrive too late and the waters are gone and the arm is already half in the birth passage or at least the hand is there. When it is early on and only the hand is out, the hand can easily be brought back inside because the child is lying with its head crooked and the birth passage is still empty. In that case you can follow the child's hand with your hand and, insofar as it is possible, assist its head as described. If the arm is already half way out, it is not possible to redirect the child's head, which is lying crooked, and you cannot put the child's arm, that is, the hand, back inside because the child's chest and torso are pushing on the arm. Nevertheless, there is still time to search for its feet with your hand and get them out. I leave the arm where it is and insert my hand as deep as necessary to search for the feet and get hold of them. The sooner this happens, the better it goes without your having to take a cord and snare the feet; at that point the child still lies loose and has enough moisture around it for it to be easy to turn it with your hand. But when its entire arm has pushed into the birth passage up to the shoulder, the child is already pressed in and in danger of losing its life. The sooner it is helped, the better it can be helped, as I show you in these four copper engravings in a case where the head goes off to one side, bringing the hand before the birth passage. Take a look at figure 2 and see how you intervene in figure 3, where you take your hand and guide the head; it shows you how to prevent the ensuing danger in a timely fashion. If you do not practice prevention, then the sort of hazard pictured in figure 4 can ensue. If it comes to such danger, you must perform the intervention I described, as depicted in figures 4 and 5. If you manage to keep the child alive and the arm is, as I said above, already well into the birth passage and the child's feet are way behind it deep inside, you cannot do anything but put a loop over the feet so you can pull on them with your left hand and govern the child and its arm with your right hand. Putting a loop around the feet does not put the child's life in danger, but its feet can be injured if you pull carelessly. Thus you must pull gradually, and your right hand must help get the child and its arm up inside so its feet have room to come out. If you are careful doing this, it will not injure the child, neither hurt its feet nor put its life in danger. As long as the child is still living, you have a little more room to guide it and it will not be necessary to use an instrument as you would were it already dead when you were summoned to assist. In that case you must undertake the extreme turn as the copper engravings of the turns will point out later.

But before this it will be necessary to show you still more arm presentations; for example, when the hands and feet come down into the birth passage at the same time, you must seize the feet quickly when the waters break on their own, that is, as Nature wills it, as illustrated in figure 6. If the hands should want to come out before the feet, you can easily put them back inside and then pull out the feet before the hands, as you see in figure 6. Otherwise, the feet may plant themselves firmly to one side and the little hands may come forth, if not both, one of them, as often happens when the child's hands and feet lie below its body toward the birth passage and the one hand comes out by itself, and during lingering labor the child gets so tightly compressed that it can and will die from it, when it otherwise could have survived. But it does not always happen, and often and usually the feet come forth before the hands. I am merely warning you because I have dealt with precisely such cases: if misfortune should come to pass, you will be informed and can be all the more careful and immediately assist with the feet when the waters break, and thus you will be guarded against it. For as long as the waters have not broken, the child has lots of room and flaps with its hands and feet beneath it, as figure 7 indicates, especially when there is lots of water around it. When the waters break and run off, the labor pains gradually squeeze in the child. If, however, you help the feet out quickly and push the hands back, the birth will ensue, though by the feet. You cannot change that, and it is impossible to get the head out first when children lie like that, as figures 6 and 7 indicate. If you brought the head before the birth passage, the child would remain curled up and could not be delivered. I tried it once. I wish never to do it again. The mother nearly lost her life. Take a look at the testimony (which is the first one) of Frau Thym.[80] I selected it as a lesson and warning to you; you will read in it that I had to push the child's head back again. If, however, the legs slip out before the arms because they are long and better able to slip out ahead than the hands, especially when both feet are together and stay together, then mother and child are not in much danger. Often those births ensue without anyone getting hurt. The births recounted here can, as I said before, to some extent be prevented, but not entirely—as can be seen in figures 2, 3, and 4. These postures have been explained because they can be prevented as with the first posture, figure 2, if the little hand should come forth before the head. First, the head lies only a little to one side. I have at various times noted that I could feel the fingers wiggling beside the head until there was room for the little hand. Then it is able to push itself way to one side, and that kind of dangerous birth results. Such things commonly happen with large amounts of water where the child has room, as well as when the skin around the waters is too tough, so the waters are too slow to break and a woman must labor unnecessarily an entire day and longer, even if the birth comes off without the child being injured. Therefore it is necessary to help with these waters so this kind of danger is prevented. For if the waters run off at the right time, you need not fear this danger. But the child must lie right; otherwise it would be irresponsible to break the waters. For inasmuch as it is possible to keep a child in the right posture by breaking the waters, even if it has slipped to one side with its little hand firmly planted, as I said before, as a result of the large amount of water or of waters surrounded by a tough skin, the waters can be broken with good

80. Siegemund refers here to the testimony included in chapter 7.

conscience, the little hand pushed back inside, and the child's head guided back to the birth passage. When the waters run off, the child's shelter is compressed. The child is caught, and a happy birth always ensues, as I have often experienced personally. Thus if the waters were to be broken in the case of children lying wrong, it would not be justifiable. But if they break naturally so it is impossible to keep them in and if the child is not lying right, it cannot turn thereafter because it is dry and pressed in, and an unnatural birth must follow. If it happens naturally, no one is at fault. I will speak further of breaking the waters and how I came to know about it. Now I will recount further how with God's help I managed difficult births at the behest of poor people. I have outlined various arm presentations I have treated, starting with my very first attempt. But I forgot to report that when the child's arm has come out up to the elbow, through the sheath and out of the woman's body, then the child is already in the mouth with its shoulder. In that case it is difficult to get it back in without sound knowledge, especially if the child's head is already bent off to one side. But the child's head does not always go off to one side, especially if it is small. With God's help such a child can often be born without injury if the head but come forth right along with the arm and stay right on or next to it. But it has got to be small. If, however, it is large, it is hard for it to be born this way. During my first delivery, when I began my apprenticeship, the child's arm was soon hanging out of the belly up to the elbow, but its head must have still lain beneath its body in the right posture for birth and must have gotten hung up and firmly planted on the sharebone, as I now can venture an opinion on it. For it was not a small child. Thus it must have been God's express will to preserve the life of the child that lay in this posture, because it seldom happens that the head remains lying in the right posture down below. When the waters have broken and the hand drops into the birth passage, no child can remain for long with its head right beneath it because it is tired out by the labor pains; so it drops its head to one side or the other, which I became aware of during the years of my apprenticeship. It was thus perhaps in this, the first delivery of my apprenticeship, God's express will to settle me in my profession and make me certain in it in order to preserve the poor woman for the sake of her dear children who needed rearing. And that is exactly why I am presenting my knowledge of this to you in detail, as much as is known to me, so you can be guided by it. To speak further of such arm presentations, I have been summoned in dangerous cases where the child had gotten fixed by the rump on the sharebone. Its little feet had gone upward toward its little neck, but the pains had forced it to present with the one hand and to one side, as figure 8 shows, and so it likewise perished. But it only lacked timely intervention, namely, the midwife must not let the rump bump up against and fix [on the sharebone—*trans.*] right at the beginning of birth and instead always guide it straight into the birth passage. You can easily feel by the child's rectum whether the hind end is coming forth straight and on which side it is disposed to get fixed. Then the midwife must introduce her fingers to divert it from the side where it is disposed to lodge itself. It will pull back its hand on its own then. But if the child remains lying crooked, it will sink more and more to one side until it dies. You cannot, to be sure, manage to get the child to be born naturally because it is impossible with such a posture. Nevertheless, you can manage a happy birth so the woman is happily delivered of the child, which thing I did after three days, and the woman survived it, healthy and happy. As experience has frequently shown me, if the midwife had done it immediately at the beginning and if only the hind end had been guided straight into the birth passage, the death of the child would have been prevented as well. And I believe these to be the happiest of wrong births, inasmuch as the child's life can easily be saved if the midwife helps the child into the birth passage as soon as the aforementioned waters break. Thus it retains its vigor and can pass through forcefully. Twins also force open the mouth of the womb so wide that the child whose head comes out thereafter passes through right away. Often such children slip out on their own, and a happy birth ensues even when the midwife has felt and noted the little hand in the birth canal and lost it again and does not know what happened to it. I have (thank God) lost few children in such births when I was there from the start. It will probably be a difficult birth for the woman but not so dangerous for mother and child. If only you know a bit about it, this one is still the easiest to assist of the births described here. Sometimes they work even without assistance or the midwife's art, if God wishes it, especially when strong constitutions are involved. Much more terrible arm presentations have, however, come under my care, for example,

When the child lies on its back athwart the womb with its head under the woman's left breast and its little feet under the right breast, the rump, however, in the woman's right side so the child lies completely backwards with the one little hand behind its back reaching into the birth passage. This is a difficult case for turning. When labor has already gone on for several days, the child is so squeezed from the force of the throes that you cannot get into the belly with a single finger. Every time I was summoned in such cases, they said, "The belly is locked up, and all the pains are gone." It was so in this one, for the belly was so bogged down from twins that everything was closed up and the pains had no more force. So in such deliveries I have had the woman lie down with her upper torso down low and her belly somewhat elevated (but not too much) and her thighs in the air with her backbone straight so the belly could get air. The more the woman curls up her body, the more space there is in the belly for me to get to her easily and the child lets itself be pushed back a little so there is some room for a hand. Then I have had to snuggle my hand underneath the entire child and its arm and pull the child's feet out from under its neck, when it was squeezed together like that, and guide it to the exit, as can be seen in the next engravings, 17, 18, 19, 20, and 21, where I illustrate the entire turn. I have had a lot of cases like that and still managed pretty well (thank God). I have dealt with such postures more often than all the previously mentioned ones, as Nature can sooner force out children in the aforementioned postures than those in this one. An experienced midwife can also certainly prevent it so it will not be so dangerous. But Nature cannot alter such a condition. Most cases where mother and child remain together[81] occur when a midwife lacks the fundamental knowledge to enable her to assist in undertaking that kind of manipulation. If a midwife with this fundamental knowledge comes up against this situation, then she has reason to intervene if the waters break since there is room enough to turn the child early on, but it is impossible to turn it by the head when the waters have already broken. As far as turning it is concerned, you must seize it by the feet—and there is time because there is still room to seize the feet with your hand—and gently bring them down beneath the child into position for birth while the waters are running off; this is completely without danger to mother and child. But this delivery works better one time than another for saving the child's life. It might also come to pass, though seldom, that when you touch the child while undertaking to turn it, it will suddenly turn over completely. There still has to be a lot of water around the child in the womb if it is to suffer itself to be turned over completely so its head with its feet behind it presents at once for birth. Nevertheless, its head remains to one side, and it is thus hard to guide it into the birth passage. It must, however, happen quickly, before the waters run off entirely; otherwise the child will be dry and will remain lying curled up, and thus you make it more hazardous than if you had kept hold of its feet—as I said before. It works better, however, if the midwife has a good understanding of turning. For as soon as labor commences, by which you must be guided, and when the waters are still intact (inasmuch as there is usually a lot of water present with those children who lie loose), if they throw themselves with their arms into the wrong posture and remain lying in the wrong posture, then they have a lot of room to throw themselves into all manner

81. That is, the child is not removed from the mother.

of wrong postures, as the case may be, as the child that lies loose in figure 9 illustrates. I have treated this kind of posture that children assume at the commencement of birth with the waters still intact where I felt the child's hand lying behind its back. So I squeezed its hand with the fingers I had introduced as if I meant to pinch it in the manner you can see in the previous engraving, figure 2, where I did precisely that so the child would pull in its hand. So I succeeded in doing it several times when the child turned itself over completely with its hands and feet and with its head beneath it, presenting for birth, so it was not necessary for me to turn it when the waters broke except to help the child through the birth passage as best I could, and several times a happy birth still ensued. When it works, it is easy to do this turn, but it does not always work; there has to be a lot of water around the child if it is to succeed. If this is not the case, the midwife, as stated above, has to introduce her hand as soon as the waters break and turn the child. If this is not done, then as the waters run off, the child's hand and arm will come before the birth passage and push into it—figure 9 shows the beginning, how the arm can sink down into the birth passage with the child's entire body descending backwards and press hard into the birth passage up to the shoulder. So you have to turn the child if the woman's life is to be saved—I will demonstrate these kinds of difficult turns to you—because at the breaking of the waters the child is squeezed so hard and the space around it becomes so small that even with great effort and great skill you can hardly get hold of its little feet (as in the aforementioned figures 17, 18, 19, 20, and 21, which depict turning the child); they are small and can be governed by the ankles, which is not possible with the head. Since the head is too large, it is also pushed in too tight, and it is impossible to dislodge it from where it lies. When you push or stroke it, it may appear to be giving way. However, as soon as you leave off stroking, it will be back where it was; it is not a movement back, but only a giving way. The child lies in the womb as in a wet cloth that clings to the child's body. So consider this then: what if I had on a wet chemise that was over my head as well and you were to pull me out of it? I will show you the reverse: pulling the chemise off me ought to be possible, while pulling me out of the chemise is certainly possible, but hard; however, it would be harder still if I were to be turned around in it. So if a child is to be turned by the head in a living belly, you can think how painful that would be; it would be impossible to bear, just as it would be impossible to budge the head from the spot on account of its being packed in tight when the waters have already run off. But in all deliveries, though it is more difficult with one than with another, you can, as I said before, get hold of the little feet because of the ankles, which are small and slender, depending on whether God lends you His blessing and strength. And thus you and I have cause to pray to God for it.

Children are also delivered by the breech or rump, and neither mother nor child are readily harmed by it. If you need be afraid of anything, it is that something might happen to the child. Of course it is hard for the woman, but not dangerous, and there is no reason to worry, whatever the midwife is like. If she is experienced, however, she can help the woman and the child a lot. See in figure 10 how she takes hold of the child by the hips with her right hand and spreads out the inward mouth with the first two fingers of her left hand by her thumb so she can work with the force that her assistance and the pains bring with them and so she can hold the child down firmly when the pains come. Thus the throes that are driven by Nature can better push the child into and through the birth passage. If the child is not held down somewhat firmly (but not so firmly as to cause it harm), it will always pull its rump up toward itself as much as it can, especially when the pains come, and it thus hinders or delays the delivery until it dies or is weakened to such an extent that it can no longer fight against it; only then does the birth take place. This resistance on the part of children readily occurs when the children are strong and large or when the woman tends to have difficulty giving birth; birth is delayed most often on account of these two causes. If, however, the child is small and the woman well disposed for giving birth, then it takes place on its own without injury and without the midwife's aid. Nevertheless, adroit assistance in such deliveries can provide all the more certainty.

CHRISTINA: It is strange that a child born rump first is more likely to survive than a child born feet first. Why is that?

JUSTINA: I have had many such deliveries under my care; at the very least it was still possible for the child to be baptized. Most of them were saved and not injured during the delivery. I lost most of those I delivered by the feet and saved only a few. So I think when the child is doubled up, it stretches the birth canal so wide that the head can get through quickly, as experience has frequently taught me. The feet are small by contrast, and the child's entire body is not as stout as the head. If this happens in the case of a belly that has difficulty giving birth and where force needs to be applied, it is impossible to save the child. If you let up on it and do not force it when the pains come, it cannot budge and thereby perishes. Even if you intend to push it when the pains come, it will not be forced because most of the child or fetus has already been born. The labor pains easily get stuck in the empty belly, and it often turns out very unhappily for the children, and most of them lose their lives.

Above it was mentioned that if the child lies athwart the womb on its back, it is difficult to turn it, especially if the child's head lies more toward the left breast or side of the woman than toward the right one. It is hard for the midwife because it is awkward for her to get at and she has to turn it against her hand; it is, however, still more difficult if no hand at all presents here, as can be seen in figure 11. The hand tells me to which side the child's head lies, and I can always get through better by slipping past the child's shoulders under the arms to search for the feet. I have been summoned several times in such situations when the woman has already been in travail for three to four days where no hand showed itself until the child's backbone was forced to break. That cannot occur, however, until it is dead and can no longer put up resistance. As long as it is still alive, it lies high up, and it is hard to reach the child or the fetus when the waters have broken, especially for inexperienced midwives. They say then that it is not yet time for the delivery, especially when there is no respectable amount of water present. Even when there is a bit of water, it trickles out. There are many deliveries where there is little—indeed hardly any—water present so you could say for sure that the waters have broken, yet you cannot tell when they break because, as I said, there is only a little present and what there is trickles out. I have been fetched various times to cases where they did not think labor had begun yet when in fact it had been going on for two, even three days. Before I could travel the several miles to get there, mother and child were dead. Because in such a circumstance the child cannot present with any part of the body, they say then, "You cannot reach anything and you cannot feel the child; it cannot be labor; it is not time yet for the delivery," and meanwhile mother and child meet their death. I have also had the experience with that sort of occurrence that I could still manage to save the mother. It is, however, very hard to turn the child in this situation and to push it back so you can find the feet. And even when I have found the feet immediately, it was impossible to pull on them with one hand so the child could turn over. So I put loops around the child's feet, as indicated in figure 11, so I could pull on the cords with my left hand. With my right hand I helped the child up and back. Thus it went happily for the woman and without injury to her; I have often tried this.

A bad birth likewise endangering the child's life is when the child presents with its belly and it is not averted when the waters break until the child is weak and dead. Figure 12 shows how to prevent it. The midwife pulls the child's head down toward the birth passage with her hand. If no one is on hand to avert this evil—for you must do it immediately when the waters break—then there is nothing left to do but turn the child, as figure 13 indicates, if the mother (or woman) is to be saved. Where the child, however, cannot be brought before and into the birth passage—since it does not always work—as can be seen in the preceding engraving, figure 12, which depicts how you prevent it, then you have to turn the child by the feet and put a loop over them so your right hand can govern the child's body and your left hand can pull the child's feet down below its body, as I demonstrated in the preceding engraving and as I will further demonstrate in the turn that follows. Here the right hand lifts up the child's body or its side, as far as possible, so its feet can get through above or below, however you can manage it best. Thus, with your right hand you have to lift up the child's belly and hips, as it lies to one side, so its feet can gradually come down below its body. I will further demonstrate to you with the following turn how it can go, inasmuch as there is a big difference in these births where you turn the child. When the child is dying or has just died, it is not pressed in as tight as it is when it has lain dead for two to three days and more, being squeezed by the pains. Early on you can often easily turn it by the feet with your hand without a cord, as I have already shown—both in this bad birth and in the other aforementioned ones—if you do it soon, since it is not possible to bring the head before the birth passage except when there is still room. In the following engraving, figure 13, however, it cannot occur without putting a loop around the child to deliver it, as happens in an unnatural birth. Nevertheless, it goes better one time than the other without the aid of ribbons. For example, if the feet lie together and the child is still somewhat wet or, as I call it, still has some sliminess around it. Before the pains have squeezed the child for several days, it always has some wetness or sliminess around it—however, one more than the other. Where there is a lot of water with a woman while she is with child, there is more moisture and sliminess with the child, though the waters break, than with women who have little water with their child while they are carrying it. Those wet and slimy children, especially when their feet are together or not far apart, are first of all very easy to turn and to get out without the help of ribbons or cords. And if the feet lie too far apart and you have to get a cord or ribbon right away to secure the first little foot you find, then you can hold this cord between your fingers when you introduce your hand, or go back and get it since you can easily fetch this cord and get your hand back in, when the child is not squeezed in so tight, as long as you know ahead of time where the little foot is. Secure the first little foot tightly with the cord then, as you see there, where the loop has been placed around the foot. Thereafter hold the cord that is around the foot firmly with your left hand so the foot cannot push forward in the belly while you are searching with your right hand for the other little foot. When you have both feet, you can pull on the foot with the loop around it by the cord from without and keep the one in your right hand in your hand and thus get both of them together and down before the birth passage. But introducing and pulling back your hand does not work so well, which I will further demonstrate to you, when the child has already lain dead for a day and more. Here the child is squeezed in so tight that you need your wits about you and the strength of your hand and arm to get in and get to the child's feet. You cannot take in a cord with you in this case since you have to search for the feet first. And even if you locate one foot right away, you cannot look for the other one until you have put a loop around the first one, unless they lie close together. And even if both feet lie close together, they have to have the loop around them before you can pull on them. The feet cannot get down below until your right hand lifts up the child's body. Then your left hand can gradually get the feet down below by pulling on the cord, as you will see in my demonstration of the turn.

Before I show you in detail how to turn a child when it is already dead and it is only a matter of rescuing the mother, I will first show you a bad arm presentation, to wit, where both hands are trying to get into the birth passage beneath the child's head (see figures 15 and 16). I cannot sufficiently describe what a terrible delivery this is for mother and child, especially when the midwife does not know what to do or is too long in getting to it when the waters have broken. If, however, the midwife is present in good time, she can easily prevent this bad birth. As long as the waters have not broken, the hands and the head remain close together so you can easily feel both hands, along with the head. If you pinch both little hands, as I already reported on pages 97–98 and as figure 2 indicates, the child will pull back its hands and the head will press into the birth passage; especially if you pinch the child's hands shortly before the throes come, the throes will push the child's head all the more forcefully into the birth passage when the child pulls back its hands. If it lies right, the head descends, moreover, while the child pulls its hands toward itself and the pains push and the delivery ensues happily and properly for mother and child. If, as often happens, the child should pull back its head as well on account of the pinching and be disposed to turn over up high, you must leave off pinching. You can soon see whether it can lift itself up high enough to turn over. If the child lifts itself up so high that you can no longer reach it, it may soon turn over; you can depend on that. It does not so readily turn over the first time; it will come back where you can reach it. Often the hands come back with it when the head descends again; often not. If they stay back, then so much the better; if they come back with the head, then you must leave off pinching and instead break the waters if they should indeed remain intact and if you can get at them. So pinch the little hands as soon as the waters run off so the child pulls them toward itself, or if the mouth is open wide enough, introduce your hand immediately with the breaking of the waters and put the hand back up under the child's head and assist the head into the birth passage; in that case you can expect a natural and happy birth. If you do not break the waters and the child has enough room (as I said) to lift itself up and back and turn over, then you can save the child's life by breaking the waters in the manner I have described to you because the child can now no longer turn. If the waters break on their own and the midwife does not put the child's arms back right away, they will soon push into the birth passage, little by little, up to the child's chest. At that point the child's neck will break, as in figure 16, which depicts the head beginning to bend backwards. The head bends as far back and up as it can so the child's throat fixes on the woman's sharebone. The weaker the child becomes and the less able to defend itself, the stronger and harder the pains push it forward in this manner. Then there is nothing else to be done but save the woman's life and turn the child by its feet. You can get in under the child's chest to the feet only with great difficulty when it is already dead because the child's chest has already pressed so far into the birth passage and its two arms, along with the chest, pretty much fill it up so the midwife can introduce her hand only with great difficulty. So you have to amputate both arms, because the child is dead anyhow. I have never found it necessary because I have small hands. But midwives with large hands—they will not get through. Thus it is better to amputate the dead child's arms because you can thus better push back the chest to get more room and rescue the mother rather than letting her die with the dead child inside. When you have gotten the child back up good and high inside, you can reach the feet, snare them, and as shown previously, pull them out, as the following engravings illustrating turning the child will also indicate, namely, how I can and must insert my arm up to the elbow to search for the feet if the child is to be turned and the woman rescued and how I have introduced in such dire emergencies a little rod, along with the cord, with my left hand because it is impossible for the woman and for me to fetch the cord or ribbon with my right hand; my left hand can, however, get it to my right hand in no other way than with a slender little rod, which intervention I have often tried. This little rod, together with the cord, and how the loop has to be pressed into it is illustrated to one side of figure 17; refer to it. There are five engravings that depict the turn up to the birth of the head. Because you and all midwives must be familiar with the sort of birth when the child comes feet first, I did not think it necessary to explain and detail it further. Nevertheless, I want to inform you about what is most necessary to be mindful of in these kinds of births, to wit, if the child is born feet first up to the neck or head and the child's chin is obstructed by or fixed on the sharebone, you must relieve it off with the customary two fingers. See figure 21. Likewise, if the inward mouth clamps down over the neck or chin, as often occurs in the same manner as with the sharebone. This obstruction or clamping down of the inward mouth occurs when the child lies with its face toward the rectum. You can, however, readily relieve it, as with the sharebone, with your fingers. Just pay attention and you will learn how to do it. Now the five engravings follow that show you the entire turning of dead children that are impossible to deliver if they are not turned. To wit, look at figure 17; it shows you, as I said before, how the midwife intervenes by introducing her entire arm, along with the little rod, into which a string has been pressed, how her left hand has to get it to her right hand, how her right hand takes hold of the loop from the little rod and removes it, and how she puts the loop round the child's foot. Look at figure 18; you will see how to snare the child and how you can do it in such postures when the child's feet are up above its body. Figure 19 shows how to intervene by turning the child, namely, how your left hand pulls on the feet, which have had loops put round them, how your right hand makes some space for the feet by pushing up the child from under its arm, and how the child is pulled out little by little. Take a look at the child's legs so you can see the woman cannot be hurried and the midwife cannot pull the feet at once with her hands. Figure 20 shows more of turning where the child is already turned over, how, by being pulled and pushed up, it gradually can yield until it is finally born up to the head, as figure 21 indicates and as I have already further described the intervention to you.

CHRISTINA: Dear God! Do tell me! Are we nearly at an end with the bad births and children lying in unnatural postures, insofar as you are familiar with them?

JUSTINA: No! I will show you still more, as for example in figure 22, where the child lies curled up with its feet down below and its face, along with its belly, facing the woman's back, which situation keeps it from turning itself naturally to a posture for birth. The midwife can aid in this delivery in no better way—although it is not easy to do it—than by guiding the feet with her hand into the birth passage so the child's feet do not fix on either side of the sharebone. However, if it does indeed occur, I want you to take note that at the very least it could cause a protracted delivery. Some children enter the birth passage by the feet on their own, as indicated in figure 23, where they lie crooked to one side. This posture does not, however, harm the delivery in any way; however much the child lies to one side or the other, if its feet but enter the birth canal straight, the birth will follow immediately as it usually does with those that come feet first. But, as shown in figure 23, if one of them should try to separate out, you must hold the two feet together; [the child—*trans.*] should go in straight and not cause a protracted delivery.

Children also tend to come before the birth passage with their knees, as shown in figure 24. In that case the midwife must take hold of the feet, as well as she can, if a protracted birth is not to ensue. I previously reminded you that when you take hold of the child's feet you ought always to turn and guide the child with its face to the woman's back if possible so it will not run up against the woman's sharebone with its chin. But it is difficult or often completely impossible to guide children, who lie with their legs down, so their face is toward the woman's back. If it is possible, however, it is better for saving the child for it not to get fixed by its chin on the sharebone. If it is not relieved off it in a timely fashion, it will die, as I have frequently noted. Figure 25 depicts twins that lie together within a single skin.[82] However, it is quite rare that there is no skin[83] in between them. In that case they cannot embrace one another as seen in this engraving; however, they can lie in all kinds of bad postures, one worse than the other. So you have to help the children in all kinds of ways, however you find them, help the one come forward and get the other moving from behind. It is like that in this situation too where both children are together in a single skin and lie embracing one another. When you take hold of one of them, the other will stay behind. So try, as soon as the waters break, to guide the one lying on top by the head before the birth passage. The delivery of the child on top will follow, and the other will stay behind until this one is born. Thereafter the other will follow feet first, and since the waters are already broken, it cannot turn thereafter. If it does not follow, then you can retrieve it with your hand, as explained above.

And thus I have shown you all the engravings, insofar as I resolved to do. To be sure there would be more of them to show you, such as children with heads that are too large and children with heads that lie crooked. Because I have already sufficiently described these, you will hopefully be able to manage without copper engravings.

CHRISTINA: Dear God! Do tell me! You have reviewed thirty-three engravings or likenesses of various births to instruct me: so which of the postures in these likenesses involved turning, or rather, which postures do you have to turn?

82. Amniotic sac.

83. Membrane.

JUSTINA: You must not yet have read my description carefully; otherwise you would not ask such a thing. I thought I had described it in such detail and so clearly that any midwife who should read this would immediately grasp it. Since, however, you cannot yet grasp it and there are probably more people like you, I will repeat it once again to inform you better.

First, where the child is already lying with its head and shoulder toward the birth passage and the child's body is up above, as figure c indicates, insofar as this posture is not immediately prevented when the waters break.

Second, when the child's right arm has entered the birth passage and its feet have lain up high under the woman's breast, as pulling them down with the cords indicates. See figures 4 and 5.

Third, when the child lies with its back to the birth canal and its arm penetrates the birth canal from behind, as figure 9 indicates.

Fourth, when its feet have a loop around them and the child lies on its back like before, only its hand has not penetrated the birth passage from behind. See figure 11.

Fifth, when the child lies belly and navel string first. See figure 13.

Sixth, when both of the child's hands are already in the birth passage and its head is pushed back against its back, as indicated in figure 16.

Seventh, look at the entire turn illustrated in five engravings, namely, 17, 18, 19, 20, and 21. The last of these depicting the turn, designated figure 21, shows how it is possible to get the child out to a certain point by turning it. All midwives know how to get the child out thereafter in such an unnatural birth. And even if some midwife were not to know about it, Nature forces it on its own by means of the pains without any special help from the midwife. All of these have to be turned if the mother's life is to be saved and all of these cannot be turned in any other way than by the feet when the waters have broken. The sooner the child is turned, the better it is for mother and child.

CHRISTINA: Why do you set the likenesses you have named apart from the others? There are certainly more unnatural births besides the ones you have just named. Is nothing to be done in the case of the others?

JUSTINA: I am showing you these up front because you asked about turning the child; the others do not need to be turned insofar as they can be delivered in those postures, though hardly without assistance. If the mother receives skilled assistance, it is all the better for her and the child. Nature can force these other births, as wrong as they may be—like those where the child presents with its feet—to occur on their own by means of violent throes, as long as the woman and child are strong. Whether the child turns its face toward the woman's back or her belly, it will be all right, though it is more difficult when its face comes toward the belly, because it can easily fix on the mother's sharebone; the child can suffocate more readily in this case than in the other birth when it faces the back. I have seen it turn out all right with still harder births, as, for example, when the child presents with its face (see figure h), though it is numbered among the impossible deliveries. Nevertheless, I have saved the life of a number of these with the help of God. And even if the child were to perish immediately, Nature forces such deliveries automatically without the aid and wits of the midwife. The woman simply has to labor long and hard, but most women survive that kind of birth.

In addition, when the little hands and feet try to come out at the same time (as indicated in figure 6), the pains usually force the feet out because they are more capable of sliding. But it does not always happen that way, for when the throes come, the child tries to get into position with its feet because it has more energy for gliding with them than with its arms. The delivery can thus finally ensue, especially when there is a lot of water and a strong woman with lots of energy. Also, if it comes knees first, Nature forces it as with the feet. Likewise, if it comes forth by the rump or if a hand and the head present at the same time and the head does not fix on the sharebone, since the fetus is very small. Also, even if a little hand comes forth along with the rump, it is not always dangerous. Furthermore, a little hand may present, as reported above, when the feet stay behind only a little. Nevertheless it is not dangerous here either.

These births fool midwives the most of all, as if all children could be delivered, however they may lie, if the appointed hour but arrive. And they have a knack for bragging a lot about such deliveries but do not know it sometimes can happen naturally on its own with violent pains and as a result of the children lying in a posture that is still conducive to delivery. Nature can often force the previously described unnatural births without the midwife's skill and knowledge, and so a happy delivery can often ensue if God wills it. Thus it often comes to pass that the ignorance of the midwife results in dead children and two or three days of labor and Nature nevertheless forces the birth, and so the midwife does not know what is going on and says then, "When the appointed hour arrives, then it happens quickly." The same thing often happens with those natural births, which certainly require assistance, as shown in figure b, also e and f and g, as well as figures h and i. But in such natural births where there are these kinds of deficiencies, a woman is forced to labor for several days, and nevertheless, after hard travail, it ends happily at least for the mother (God willing). So the midwife deceives herself into believing all children can be delivered without any special intervention if the hour for delivery has but arrived. They do know how to force the pains, and it is for this reason that they are so eager to force them. And they do it indiscriminately and without proper cause.

I hope you have understood the information I have provided about assisting birth as I have come to know it and that frequent practice will enable you to decide how it is best done; God's blessing and good sense must do their best. That is why I have described touching or manual examination to you so precisely and truly. You will thereby come to know what skilled assistance is, something hidden to many midwives.

As far as twins are concerned, you have to deal with them in many ways, as they lie, by turning them, or with skillful assistance; just as with the many previously cited deliveries, you have to be guided by how they lie, with one and then the other. Two little children can be better managed than many a large one, which reason should tell you. And I have already explained with the same engraving how to assist them. Do read it carefully.

CHRISTINA: Dear sister, the first postures you cite, along with the ways to turn them, are terrible. Can you not provide assistance early on to prevent them so you need not turn them in such an extreme fashion?

JUSTINA: Yes, you can pretty much prevent them, if not entirely. When I have been present from the start, I did not let it come to such deliveries. Touching and the child's posture will inform you when the waters are still intact which posture can become bad like that. You can assist the child best with the waters still intact. I will describe it to you one more time, though I have already sufficiently reported on how, with God's help, I prevented it.

First, when the child's head lies crooked (as can be seen in figure c). It does not lie that way when the waters are still intact; the child is still free then to have its head crooked or straight. Nevertheless, you can tell whenever it tosses it or fiddles around with it. Every time the child moves, you feel either the open head[84] or an ear, but it lies very high up. That gives you a lot of information because it tends to change often. In such circumstances I have watched for when the open head turned down below and when true pains, as well as a proper opening of the mouth of the womb, were present. Then I broke the waters and helped the head into place, and felicitous deliveries ensued. For as long as the waters are intact, the head does not lie so very crooked before the birth passage. But if the waters break and it happens with the head lying quite crooked, mother and child are in danger of losing their lives. Thereafter it can no longer help itself and straighten out its head, and from one pain to the next it will be forced into a worse and worse pos-

84. Fontanel.

ture until it suffocates. I have been summoned in the case of such a posture several times where the child was just barely alive, yet I have been unable to assist in bringing the head into the proper posture because the waters had broken and run off. In the end, I could only save the mother by turning the child, and I had to seize it by the feet if she was to survive.

Regarding the second of these seven, when the child presents with its arm, as indicated in figure 4, you can assist and prevent the danger in this birth, if not entirely then largely, when the waters are still intact right when the true pains commence. In general these children initially lie right and have their hand by their head—I already told you this and also made clear to you how to prevent it, as I have often found and tried it. If, however, they are lying this way and the waters break with a good opening of the inward mouth, you must introduce your hand rapidly at the breaking of the waters, for I have thereby succeeded many times in saving mother and child. I have brought the child's head into the right posture while the waters were still running off. If, however, the waters have already run off too much or Nature did not provide much water to start with, it is not possible to bring the malpresenting head before the birth passage (which introducing your hand and touching will soon show you). So I took hold of the feet immediately while it was roomy (I say roomy because the longer you wait, the tighter it becomes), and mother and child were often saved—if not the child, then the mother—and lingering labor was thereby prevented. I have been called in when such danger arose when nothing else could be done but rescue the mother. If the woman's strength had been sapped and she had had to lie there too long before she received proper assistance, she died anyway. So be mindful of this in such deliveries; you will always be able to help the woman if you do it in a timely fashion.

The third of the seven births reported on above is when the child presents with its back and its arm penetrates the birth passage behind its back. This is depicted in figure 9. The child has its arm behind or beneath it to one side with the waters still intact, and you will be able to feel it with the waters still intact. This has often deceived me, and I thought the child lay belly first because I could feel a hand twitching. When the waters broke, I reached for the hand immediately against the waters and found the child on its back. So I let go of the hand and helped the child upwards by its back in order to turn it over. If that did not work, then I took hold of it by the feet, and as the waters ran off, easily turned the child. Thus this danger was averted.

In the fourth of the seven births described above, when the feet have a loop put round them and the child presents with its back, but the hand is not behind it, as shown in figure 11, the child lies deep inside and is hard to reach. Because it has to be reachable when true labor pains are present in order to avert the danger, even if it is hard for the woman and the midwife, it is nevertheless even more difficult for both when the danger of the waters breaking arises and the infant remains in that posture; in that case there is nothing else to do but save the woman's life and turn the child in the manner previously demonstrated. This is much harder when the waters have broken than when the waters are intact. I have already shown you both ways of turning the child when the waters are unbroken. Just look back.

The fifth of the seven postures described above is when the child presents with its belly and the navel string (see figure 13). When the waters are still intact, the posture is not like this, but rather as is shown in figure 12. This sort of bad birth has suddenly unfolded beneath my very hands when the waters broke. When the waters were still intact, I felt the child's knees in the woman's side at the hip; on the opposite side I found the child's head lying face down. So I tried to think and deliberate as to what I should do. While I hesitated, the waters broke. When I tried to touch her again, I ran into the child's stomach or belly along with the navel string, though still deep and high in the belly; nevertheless, its head was bent back. However, because I did not know any better, the throes gradually forced the child into the birth passage. I thus had to slip my hand beneath and past the child. I got to the feet and pulled them down in front of the child into the birth passage so an unnatural birth did indeed ensue, yet mother and child were parted from one another. Since that time I have been immediately wary of that kind of posture and have paid better attention to the breaking of the waters; to wit, I have kept the customary two fingers inside the woman until the waters broke. At this moment I have introduced my entire hand, which could help to hold back the waters so they would not run off so quickly. And I have taken hold of the child's head and pulled it down. See figure 12. When I have gotten it before the birth passage, I have gradually let go of the waters, which waters guided the child still more toward and into the birth passage so the delivery ensued happily. That it why it is better to avert than to wait for this kind of danger. If you fail to learn to avert it in this way, you will be even less likely to learn how to turn the child, for turning the child is harder in this circumstance than prevention. When you are averting danger, the child still lies high up in the hollow of the belly and the waters are still there so you can get at it more easily; it is not lying crammed in as tight as when the waters have run off and it pushes farther into the birth passage. You can be guided by this.

The sixth of the aforementioned seven births shows both hands in the birth passage and the head lying back on the child's back; see figure 16, which depicts this kind of posture. This is easy to prevent at the beginning of delivery; before the waters break, the child lies as depicted in the previous figure 15 and has both hands at its head below its forehead before the birth passage. This is quite easy to feel when you touch the woman with the customary two fingers. If you pinch the little hand (as I already explained with figure 2, page 97), the child will pull in its hand and the head will slip off the sharebone, where it is lodged or pushing against it, and into the birth passage. If it, however, does not work, you must quickly intervene when the waters break and push the child's hands back up inside and pull on the head; the danger is thus averted. If you come too late and find things very bad, you have to resort to turning the child, as I demonstrated when I described this figure in detail.

The seventh and last posture I have promised to explain once again is outlined in the copper engravings depicting turning. I am of the opinion that, with God's help, all of these bad postures can be prevented. If, however, the waters break before the midwife gets there, that kind of posture ensues, and there is nothing else to do but turn the child. Likewise, if the waters break early on before the inward mouth has opened up sufficiently (it often comes to pass that the waters break three days and more before the birth), it is then also not possible to prevent. Thus you have to see to it that you turn it as soon as possible. The sooner you do it, the better it goes; the farther it penetrates into the birth passage, the harder it is to get it back inside. I have experienced great difficulty with it when I was summoned several miles away when the woman had already lain for three to four days in such travail. I cannot describe what a great effort it cost me. Nevertheless, God always aided me and granted me the ability to part mother and child. May Our Faithful God continue to provide succor! Many mothers survived and went on to live healthy lives for many years. Some, however, whose strength had been completely sapped by this lingering labor, died.

CHRISTINA: Do explain to me the births that are called natural, since you say they are also subject to complications and require skilled assistance.

JUSTINA: The first engraving illustrating these natural births depicts a child lying crooked, which with skilled assistance can easily be guided to a natural birth. See figure b.

The other engraving illustrating these so-called natural births depicts a child lying in the right posture with no impediment to delivery. See figure d.

The third, designated figure e, is a natural birth that need not be altered. But, like the previous one, it is hard to deliver because the infant is lying on its back. However, there is nothing more to be done than guide the head into the birth passage.

Figure f depicts the fourth type where children who lie right for birth require assistance. I was afraid the child's shoulders were bigger than its head, and I must thank God that I have always been lucky in that kind of circumstance, if I was but promptly summoned to the woman. I have always refrained from pushing too early and too hard, which, I have found, is the greatest help. For the less you push with a child with a large head, the better the outcome. (I am talking about pushing the pains or using medicaments early on and not about necessary or natural assistance.) The difference between this proper and able assistance and unnecessary or useless aid is obscure, and few midwives pay heed to it. I can say with the truth of experience that the gravest of the dangers that came into my care when I was summoned to assist were when the child lay right and either had too large a head or shoulders that were too broad. Because midwives do not pay proper heed to this when labor begins and are only concerned with pushing the pains and urging the women to push harder, mother and child are put in peril of their lives.

CHRISTINA: This report seems strange to me. You do not think much of vigorous strengthening of the pains or intervention in such cases when the child has a large head or broad shoulders. How can a stout child be delivered without assistance and strong pains, and what, then, should I understand our task to be?

JUSTINA: This art consists in the midwife's paying careful attention when she examines the woman to determine whether the natural pains will force out the child or not, and if it does not enter the birth passage, what is to blame. For when a child lies right and is not fixed anywhere, it has to enter the birth passage or there has to be a cause for it: either the inward mouth is unable to yield naturally (as was sufficiently recounted above), or the child's head or shoulders are too large. If one of these causes, whichever it may be, is at fault, you must not urge the woman to push harder; rather you must let the natural pains have their way and the woman must be guided by their urging or incitement, which you will perceive better than the woman in labor when you touch her. You need to govern and guide the woman according to this so she does not neglect the pains but also does not push too hard. For if you do not aid the pains and suppress them with such stout children, mother and child will remain together. On the other hand, if you push too hard, it will go just as badly and worse. So you must go the middle way. When you touch her, you can perceive quite well whether the child's shoulders are too broad, for in this case the child appears not to be crammed in as tight as when the head is too big, where there is no more room to get a somewhat deep grip on the head. When, however, the shoulders are too broad, you can get hold of the head down to the ears, but the delivery will still be slow and hard. In such deliveries I have proceeded as gently as I possibly could until the child passed through; the child thereby flexed its shoulders and head and at the same time penetrated and passed through the birth passage. For if you push too hard, the child will push its head and neck through, but the shoulders will get stuck and not be able to penetrate or push through. The child often suffocates unless the midwife be experienced enough to introduce two fingers of both hands alongside the child's neck, as soon as the head emerges, and take hold of the child by both shoulders and help it in and out.

The fifth type has great need of this last-described intervention if the child presents for birth off to one side. When one shoulder bumps up against the sharebone and fixes on it, you can relieve it off by means of the intervention I just told you about. Figure g depicts this. I have been summoned to such deliveries where the child's head had been born for over twenty-four hours and where it could not come out farther. The child was dead, and the mother sapped of all her strength, when nothing was lacking but this guidance. Whenever I guided the child, the birth followed soon thereafter. People used to say the womb had closed round the child's neck, when it was certainly not true. A terrible mistake on the part of the midwife is at fault in this closing of the womb, and it is more an excuse for her ignorance than the truth. I do not want to argue with any midwife on this account, as I am as a rule disinclined to dispute anything; may each be left to her own opinion. If, however, you understand my report and pay good heed to it, you will get to the bottom of it.

The sixth example of a head in the right posture, as shown in figure h, is when the child's face is turned the wrong way and the child has to be born face first. It is certainly necessary to provide assistance here, but it also works, God willing, without assistance, if the woman is hale and hardy, as I recounted when discussing this copper engraving.

The seventh and last of the deliveries you asked about, those when the child is said to lie right and nevertheless complications arise, concerns a child, as you can see in figure i, that lies right, but nevertheless the navel string comes forth in front of its head. It requires assistance if it is to survive—I will tell you in detail how I deal with this in the conversation about the necessity of breaking the waters. So I hope you have clearly understood my explanation about or concerning turning the child, as well as so-called natural births, and have likewise grasped the interventions that are highly necessary in all kinds of postures children lie in. I have described and also illustrated these for you with copper engravings so you will be able to save many children and women if you but familiarize yourself properly with all this and so they will not be so readily in danger of losing life and limb as often tends to happen with inexperienced midwives.

And this is everything about turning the child and providing skillful assistance I wanted to talk to you about, from the bottom up, as I have come to know about it through my own experience.

CHAPTER 5

Of those prolonged deliveries when the children lie right and yet have died in the womb; how to get them out with a crotchet when they have gotten pushed far forward and squeezed in very tight

CHRISTINA: What do I understand you to mean when you say you have been lucky in all your natural deliveries, when you also say your most dangerous cases have been this sort of delivery.

JUSTINA: As I said before, both things are true. I have indeed been successful when I have been summoned in a timely fashion. At the very least the mother recovered happily and in good health without any brutal intervention and without the crotchet or some other instrument. I do not usually have need of the crotchet if I have attended the birth from the beginning. Yet there are often complications with such bellies where the children are large and the women are up in age or stout. In such cases it is terribly difficult if they are giving birth for the first time. There is flooding, dire extremity, fainting, and the children perish in the process, especially if the delivery is managed badly. And even if it is managed properly, it is not always possible to prevent these things. Children often suffocate in these hard births. So even when I have been in attendance from the very beginning, I have had to use the crotchet to save the woman's life. But such births do not occur often if you but manage them carefully from beginning to end; whether the infants lie right or wrong, they can usually be managed without the crotchet if you turn or guide them properly. So I can say I have had little need of the crotchet if I was but there from the beginning. But in those cases when I am called in to assist where the woman is weakened from lingering labor and a hard delivery, I have found the crotchet necessary. I can truly say I have not had as much trouble with any delivery and never lost so many women in all my deliveries as when I have been called in too late. Since the children lay right, the midwives always argued against me, insisting that the children could be delivered and that they probably were still alive. During such quarrels women have languished unto death without succor. When I am finally permitted to help, it is often already over with the mother, as well as the child. So I have had to pull the infant out of the mother with a crotchet. Nevertheless I have done it without injuring her. Some of these mothers, however, died thereafter from being too fatigued and from losing their vital forces in this lingering labor. When a child lies right, it is of course difficult to know whether it is dead. So women are quite often neglected.

CHRISTINA: Would it not be justified to save the life of the mother by employing a crotchet or some other instrument in a timely fashion before all the vital forces had been sapped from her, if a child, were, contrary to expectation yet unbeknownst, still alive and yet, by all appearances, could not be delivered?

JUSTINA: This question is too lofty for me and a matter for learned men to address. I myself would like to have their answer, for in these births most women are in danger of losing their lives because they waited too long to summon me. Even though I always get the child out of the mother's belly without injuring her (which I have always done successfully whenever I have gone to work on the mother), some women, who are too sapped of their vitality, die nonetheless. It would be important to our profession to ask this question, and I would like to have legal scholars answer it. I have for a long time thought about asking it myself. But the persecution I have encountered prevented me from doing so; I was accused of wanting to remove children prematurely from women using brute force. So I refrained from posing this question, and women have thus on various occasions been neglected. I cannot refrain completely from using the crochet despite evil gossip and for this reason: I am often summoned to assist when the infant is lying right but the attending midwives have pushed it and squeezed it in hard and tight through excessive forcing of the pains during four or more days of lingering labor. Thus the child suffocates, and the woman is robbed of all her strength. In such cases you cannot get the child out in any way but with the crotchet. I wish I would never in all the days of my life meet up with or be summoned under such circumstances, because in such cases it is harder to get the child and harder to save the woman than with many a turning. As long as no one perceives the danger the woman is in, no one helps her. Thereafter, when the woman's powers fail to the point of no recovery, they all begin begging me to save her for the love of God. What should I do then? You cannot grasp the child's head with your hand, and moreover, since many a woman, who is strong in body, has been saved, you always hope it will be the case with this one too. For this reason my conscience will not allow me in cases of dire extremity to forgo the crotchet. If all midwives knew from the beginning how to manage mother and child correctly during the delivery, then it would be easy to refrain from using an instrument or crotchet. I seldom need the crotchet, if only, as I said before, I am present from the beginning of the birth on. I hope Our Dear Lord will continue to preserve me. Yet I have had to use the crotchet a number of times when the woman could not endure protracted labor on account of other complications, like copious flooding or great fatigue, and the children were dead, and so there was no other way to save the mother, though I know full well how to guide the child into the birth passage. It is indeed a great help, but such dangerous complications do not allow you much time. Soon it is a matter of life and death. However, since such complications are not common, I repeat and say once more that I can easily avoid using a crotchet or some other instrument if I am fetched at the start of the woman's labor, for I know full well how to guide the child. Proper guidance of the child and proper management of the delivery, as well as all manner of turning, are easier on mother and child and on me too than saving a woman with the crotchet or some other instrument. Why do midwives so often let it come to such dire straits? Indeed, it is usually possible to prevent it, and it is better to do something at the beginning of labor in the roomy belly than afterwards in the narrow birth passage when the waters and everything are gone.

CHRISTINA: But what happens then when you, as I know you are, are called in to assist and you nevertheless sit there a whole day and longer before mother and child are parted because you are sure to be able to assist by guiding the child, but thereafter you use the crotchet all the same?

JUSTINA: You need to know I am never summoned except in dire extremity. Indeed, I wish they would summon me in a timely fashion. When they see the birth is taking a while, and the child lies deep inside, and the pains are small or false (as they are accustomed to calling it) and are not forcing the child down, it would be time, for the child is still lusty and the mother still has her strength. I could assist them more easily and better then than later on when they have been sapped of their strength. Often you would not even have to think of using the crotchet, let alone actually use it. The difference in whether or not I have to use the crotchet is this: one woman has more strength than another to labor. A strong woman can endure longer in difficult labor than a delicate and weak one. So in one situation I am summoned later than in another. Sometimes they still have some strength left, sometimes none at all, sometimes a little. The sooner I am summoned and the more strength remaining, the more certainly and better I can assist without using the crotchet. Often, however, I will spend an entire day hoping to get the child out without using the crotchet and still have to use it in the end. This happens because I tend not to use the crotchet unless it be a matter of dire extremity. It often works without the crotchet, but takes more time and is difficult, inasmuch as it takes more time and is harder to deliver a dead child that is compressed than it would have been at the beginning of the birth. During such prolonged and difficult labor a woman is sapped of her strength before you know it. So, though I did not want to, I have to take up the crotchet to save the woman. Using the crotchet is a poor amusement for me and for the woman. But if the situation persists, there is no other way.

CHRISTINA: Why did you let in the French barber-surgeon in the case of the cooper's wife? Could you not have assisted without him so you would not have been subjected to the nasty rumor that you were not able to do anything without the crotchet and that he could have readily provided assistance as soon as he got there without the crotchet or some instrument?

JUSTINA: I called in the French barber-surgeon in the case of the cooper's wife for no other reason than to refute the accusation that I use the crotchet in all difficult births, when there is no need, and to prove that it was the slander of my enemies and that I do not use the crotchet except in dire extremity. I was summoned to this cooper's wife after she had already labored hard for two days and two nights. I found the woman suffering considerable pains with the child lying right, but already dead. It was this woman's first birth, and she was no longer young. The child was large for a first child and pushed far in so it could be moved neither forwards nor backwards. I worked an entire night, applying all my knowledge, industry, and strength, but all my efforts were in vain, for it was not possible to get hold of the head with the palms of my hand or my fingers. My fingers were too weak and slipped off the head, and my hand is of even less use with children that lie right when they are pushed in so tight and stuck because of the narrowness and tightness in the birth passage—unless you wished to proceed roughly and do violence to the mother's health. I have to confess that I much prefer to assist when the child lies wrong than when labor is prolonged like that. For if the child comes wrong and you have to turn it, the crotchet is not necessary. But to my knowledge, these children who present naturally with the head and get stuck can be drawn out only by using a hook to pull on them when the mother has already been sapped of her strength and is at death's door. So in the case of this cooper's wife, I saw the danger and most dire extremity but knew of no other way to save her than with the crotchet. So I had the venerable doctor summoned to me wondering whether he might know of another means of helping her or whether he might not want to get this barber-surgeon to come to us. If he knew of another means of helping her, I would gladly take it up and make use of it in the future. And that is how he came to us. He was immediately prepared to assist her without an instrument or crotchet and promised to have the child in an hour. It had just turned eleven in the morning when he sat down with the woman. For half a day he assiduously made every effort with the mother and child. His efforts were in vain because the child, which had been squeezed into the birth passage, was not to be budged. Along toward evening he asked me in secret, "What kind of instrument do you use in such cases, forceps[85] or a crotchet." I answered that I had nothing but a crotchet. Whereupon he continued, "If you have got it with you, let me see it. If I could make a hole in the child's head with it, I could get hold of it and pull." So I gave him the crotchet. When he tried to insert it, the woman felt it, and the crotchet was revealed. To be sure he attempted many times to insert it, but it was too large for him (as he said to me publicly in the presence of the venerable doctors and a barber-surgeon from here, as well as several women who were present) and he could not insert it on account of its size. So he asked me to sit down and try and see whether I could get it in and could make a hole in the child's head with it so the woman could be saved, because there was no other way but to hook the child and pull. I was reluctant to let myself be persuaded to do it. Nevertheless, I finally sat down in his place and found the child's head in the same place it had been when I turned the case over to him—except that the skull was now completely broken to pieces as a result of his many attempts to get hold of it (as was to be expected). Because this woman's strength could still endure long enough for the child's head to be broken into shards, it was easy to make a hole without the crotchet, which I did with my fingers. And I was surprised that this Frenchman turned this woman back over to me saying, "Since you are more accustomed to your crotchet than I am." And thus in the presence of all those in attendance my innocence was revealed: he simply knew of no other means with such a birth but to make a hole in the child's head with a crotchet or forceps so he could get hold of it and pull. Was then his knowledge herein different from mine? Indeed, it was a proof of my innocence.

CHRISTINA: In the houses of persons of quality you hear it said, nevertheless, that you could not provide help, but that this Frenchman could have offered help without the crotchet.

JUSTINA: You must know that this comes from enemies and slanderers who carried it into these homes. If I should hear this, I would show them quite the contrary through the people who were there and who would tes-

85. This instrument should not be confused with the midwifery forceps. When Siegemund refers to forceps, she refers to a tool something like tongs with pointed ends.

tify under oath that it was not otherwise than I have recounted it here. Indeed, this French barber-surgeon cannot say anything different insofar as he is as upright and conscientious as I take him to be.

CHRISTINA: But it is true, and you confess it yourself, that a crotchet was not used and you made a hole in the child's head with your fingers.

JUSTINA: It is true that no crotchet was inserted, but it was my singular good fortune that it involved a strong woman whose strength held out. Not one in ten, probably not one in twenty women would have endured had she undergone such a trial as did this woman. She had been in travail for two days and two nights when I came to her; her child was already dead—so thought the first midwife and the women who were present. On top of that, because her pains were still strong when I arrived, I continued all night long with all my might to try and assist her. Yet the child did not budge an inch, neither backwards nor forwards, and I deceived myself with good hope the whole night long, as did that Frenchman the following day. If this woman had not been so strong, she would have died on us. When the barber-surgeon arrived with new energy and worked for over half a day with his strong hands and fingers to pull and push the head, the skull finally had to break so neither the crotchet nor any other instrument was necessary. And because without a doubt God wished to bring my innocence to light, it had to be turned back over to me, inasmuch as the barber-surgeon had not dared to make a hole in its head without a crotchet. When, however, I examined it again and found the skull had been broken in two and I could get a deeper grip on the head with my tiny hand than he could with his big one, it was quite easy to make a hole with my fingers. So when I made a hole through the skin and skull, the skin around the brain was very tough. With the permission of the doctor and the others present I pierced it with a hairpin. The brains ran out then, and it became possible to take hold of the head and pull on it, and it could thus be delivered. When the Frenchman saw this, he wanted me out of the way. He said given his strength he could assist better than I, even though neither strength nor art was necessary then. Nevertheless, for the sake of peace I ceded my place to him. The moment had arrived. Hardly did he take hold of it and pull, when it was in his hands. This woman survived her travail and retained her health as well, though she was very weak. So she lived four more weeks and in the end died suddenly and unexpectedly. To be sure she was still weak. Nevertheless, she got out of bed now and then and sat up for an hour or more. And when she got up one morning to sit for a while and had been out of bed for just a little, before her mother could make up the bed for her, she cried out, "Alas, Mother, what's wrong with me!" Thereupon her mother put her back to bed. Hardly had she been brought to bed, when she lay down and died. It appeared she had suffered apoplexy.

CHRISTINA: I can well imagine this difficult birth had so affected this woman's head that she was unable to recover completely from it until finally apoplexy set in and ended her life. Thus it is a terrible thing when a mistake is made at the beginning of the delivery, as often comes to pass as a result of the ignorance of the midwives or the stubbornness of the woman in travail.

JUSTINA: Since I promised to reveal to you all I know and you remind me now of the ignorance of many midwives, as well as of women's stubbornness, let me tell you about a dangerous and difficult birth that came under my care in the last few days so you can prevent this sort of thing; the pure stubbornness of the woman in travail and the ignorance or mistake of the midwife had resulted in its becoming so dangerous and difficult. Since this intervention or turn almost seems harder to me than all the other turns I have shown you, I think it necessary to present it to you for the sake of the common weal. I found two causes or mistakes for this delivery being delayed and mishandled, and it is highly necessary for all women in travail and all midwives to know about them so they can learn to beware of this kind of folly and stubbornness on both sides. The first cause and mistake was that the complaining woman would not suffer the midwife to touch her before she was in the throes of true labor, whereas the woman's waters had broken before the midwife had been summoned. The other mistake and cause was that the midwife did not want to contradict the woman and thus stayed with her for three days and nights without touching her, waiting for the true throes and the appointed hour of birth. Because they both were of the opinion that the waters often break three days and more before the appointed hour of birth without doing any harm, they thought it would be like that this time too. But their opinion was wrongheaded, and both of them were mistaken, and so mother and child were in danger of losing their lives. It is true that the waters often break without labor pains and that not until a few days later does the happy delivery ensue. The midwife must be able to recognize the difference by touching the woman in a timely fashion and seeing whether, when the waters break, the child is lying right and whether the inward mouth is open. Some women give birth easily, and the mouth opens without much pain from carrying the child and with small pains. This is a great help to women when they deliver, in contrast to others who experience such hard deliveries. And a midwife cannot know this simply by looking. If the child lies right for birth and the mouth is already open, but with no labor pains or small ones, and even if the mouth is not open very wide and there are no pains or small ones, if the child lies right, the midwife can very well wait for the proper bearing pains before urging her to push; otherwise she would be sapped of her strength too early on before it was necessary. If the woman is not to be endangered, the midwife really must touch her early on. It will reveal the reason for the condition. This should serve you and women like that as a warning. All midwives should mark it well so they are not intimidated at such times by ignorant women like that, but instead say what is necessary to say to them. For when danger seizes mother and child and delivers them unto death, those left behind scream about the midwife, accusing her of neglecting to probe the reason for the condition, which is highly necessary. It would have gone just like that for this woman, had she not immediately been aided by having the child turned. For the infant was not only dead, but so swollen and squeezed as a result of the force of the throes and the running off of the waters that I had more difficulty with this turn than with all of the other turns I have illustrated with the copper engravings because I had a lot of trouble getting or reaching even one of the legs on account of the child being dry and prodigiously swollen. It was impossible to bring both of the legs together, and for this reason turning it was much harder than in the illustrations of turns I have already shown you.

CHRISTINA: Do not lose track of what you are saying. I need to ask you something first, because you say when breaking the waters like that the midwife needs to know how to tell whether the child lies right and whether the mouth is completely open or not. And you continue, saying if the child lies right, you should wait for proper bearing pains before urging the woman to push. You do not say, however, what I should do if the child lies wrong. So I need to know what should be done here too.

JUSTINA: When the child lies wrong and the waters are already broken, but the mouth is open properly so you can get hold of the child, you must, before the true throes come, turn it right away and bring it into a position that makes delivery possible, as I have already shown you in all manner of ways. I call them the proper, that is, the so-called bearing pains. These strong pains push the child hard and as far into the birth passage as they can, and yet ignorant midwives still do not call them the proper pains because the child cannot break through and be born. So turning the child is hard indeed, and for this reason you should turn it before these pains come. For the tolerable pains that precede these cause the mouth to open, especially with bellies that otherwise are disposed to give birth easily when the children lie right. This woman had had six or seven such easy births and was counting on that this time too. But it does not come off like that when the children lie wrong. For this reason it is better to turn the child. Because it still has a lot of room, it is easier for you and the woman to turn it then. And once it is turned, if the throes do not follow, you can wait for them, and you should not encourage the woman to push until the pains set in naturally. And you most certainly must not give her anything to force the throes. But you can certainly give her something to strengthen mother and child since the turn has tired them out. For when woman and child gain strength, the throes will set in on their own and the delivery will follow. Proper pains will readily ensue after the child is turned if it is directed toward the exit so it can enter the birth passage and if only the woman's strength is not too far gone. And the pains are delayed for precisely this reason, especially if the woman is weak of constitution or weakened from pushing a child that lies wrong, as you can see from the delivery I have just reported on. I have explained this delivery to you to enlighten you. If, however, the mouth does not open at all when the waters break and the child lies in the wrong posture, you have to wait for the mouth to open to get into the woman's belly to turn the child. The sooner you can do it, the better and easier it is to turn the child. You can help the opening along if you introduce the customary two fingers into the mouth, and when the throes come, the mouth can be widened by the fingers and be brought sooner to a full opening than it would on its own. For because the child is lying wrong, it cannot press down and get at the mouth and force it to open as it does when it is in the right posture or when the waters are intact and they force the opening because they sink with their skin into the inward mouth. And when the pains come and the water expands, it forces the mouth to open. This does not happen if the waters break so early on without labor pains. Thus your reasonable intervention can accomplish a lot so you can turn the child and the woman need not suffer for so many days or be in danger of losing her life. This warning and instruction must not be dismissed; if you pay good heed to it, you will find a good way of helping.

CHRISTINA: I admonished you earlier not to forget what you were saying because I had more questions for you, namely, how should I understand what you are saying? First you say the cause or error was that the woman in travail would not suffer the midwife to touch her until the proper bearing pains came and the midwife let this woman have her way and yet remained with her until the third day when they summoned you to assist. And yet you say you found the child swollen and squeezed by the force of the throes and the running off of the waters. How could the throes have squeezed and pushed the child forward if there were no bearing pains? Where does the kind of danger for mother and child you indicate here come from? The woman will have been forced by the throes to say something about it on her own accord, and the midwife will certainly not have been blind so as not to see that the woman was having those kinds of pains, for proper bearing pains cannot be hidden and stifled. I cannot make sense of what you are saying.

JUSTINA: I will show you the ignorance from which this danger arises so you can be on your guard and so can these women who suffer from the delusion that the midwife's touch is unnecessary prior to the most violent bearing pains. You need to know that with children who lie wrong, as I earlier related at length, as, for example, when the child lies athwart the womb, there can be no bearing pains. Just consider rightly what bearing pains are. They are nothing but the force of Nature compelling the child to exit when the womb wishes to rid itself of it, as in the natural hour of birth when the child lies right and thus is in a position to be forced out. If it does not lie right and lies so wrong that it cannot descend, the force must let up because the child cannot descend on account of lying athwart the womb. And it cannot be delivered lying athwart the womb. Position yourself in a window so as to look out. If you are positioned square in the window, someone can easily throw you out from behind. If you are positioned athwart the window or are braced crooked to one side, that person will have to leave off trying to push you out with that sort of light pushing. It is like that with the bearing pains as well. If the child lies right, the bearing pains will ensue not merely as a result of the inner force of the pains but also from the child coming into position and pushing through. When these two aspects come together, the delivery ensues without any midwives' help. So if you and those women of yours with that mind-set determine to await the bearing pains without a proper examination, you can easily be imperiled as happened with this woman. Thus these are the bearing pains when the child can penetrate the birth passage properly and be born straight away. If it lies wrong, you cannot recognize the strong pains by looking, and a woman and a midwife can thereby easily be mistaken.

CHRISTINA: Have you never before encountered such a condition other than this time? For you say that turning this child was much harder for you than those turns in all your previous copper engravings. Why do you not show me this delivery with a copper engraving too? I can grasp it better by looking at a copper engraving together with a detailed report than from the report alone. The copper engravings light up my eyes as it were and place understanding in my hands; so do show me this turn just as you did the others, or particularly, what all it takes to do this turn.

JUSTINA: You astonish me, but still I have to oblige you. Look at this first engraving [in the third and last series of engravings beginning on page 131—*trans.*] which shows you a child lying closed in quite tight and

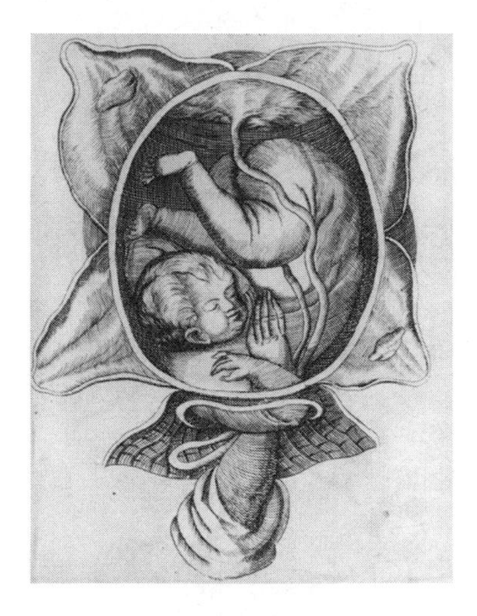

Figure 1

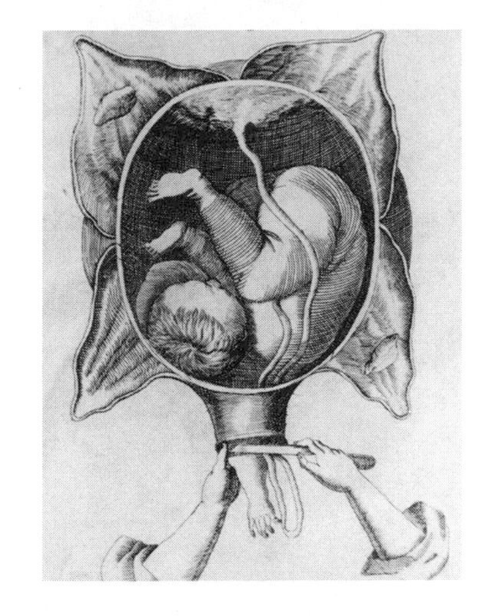

Figure 2

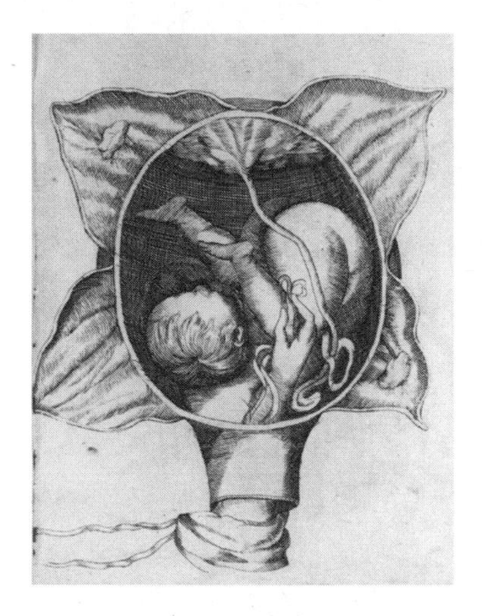

Figure 3

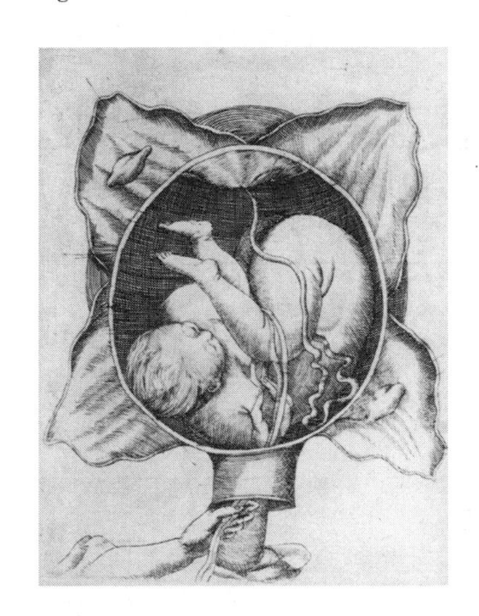

Figure 4

how I found it large and swollen and squeezed by the pains with all the waters run off too, and how it was impossible for me to push the child up high enough so I could get my arm far enough inside even to get to the nearest foot or the child's knee, just as you see here how my arm is swollen up large. For this reason, I was forced to decide to pull out my hand along with the child's arm and hand so the child could shift back closer to the exit. Because my arm had gotten pushed way up high on account of the child's squeezed-

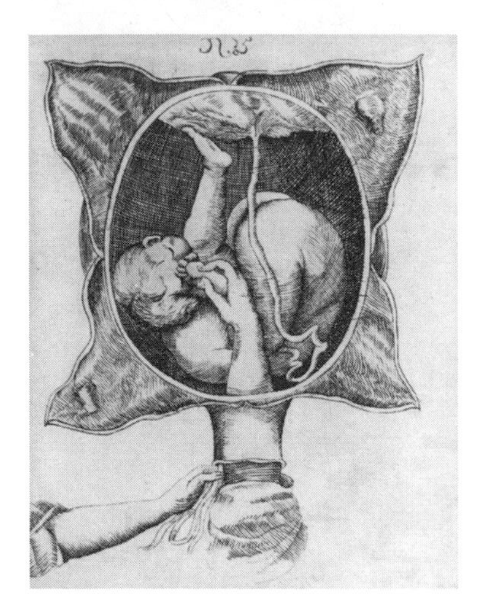

Figure 5

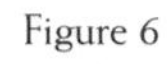

Figure 6

Figure 7

in arm and its chest, I cut off the child's arm, since it was dead anyway—as this second engraving shows—and I could not do anything about it, and since the child was stuck quite fast, I did not have to be afraid it would be pushed forth or pull back, as indeed happens if infants still have liquid around them and are not squeezed in so tight. I already showed you that sort of pulling back and that it is not always good to cut off the arm. Here, however, Necessity demanded it so I could get more room to reach in farther because the child was squeezed in so tight. The third and fourth figures show how I could and had to thread a ribbon through the leg at the knee because I could not reach or get the child until I could pull with my left hand on the leg by the ribbon I had looped around it so my right hand reached the foot, as you can see in the fifth figure. The sixth engraving shows how such a child must be compressed, since you cannot really get to it by touching. For you cannot possibly reach in with both hands, and one hand is not enough to push up the child and pull on it at the same time until you get hold of a little foot. But in turning this child I could not get hold of a foot until, as you see, I brought the foot out into the birth passage so I could put a loop around it. Then I could insert my right hand next to the child's legs in order to push it by the shoulders and belly back up high—as you can see in the other turn in figure 19. Here I seize the child under the arms in order to push it up to be able to turn it. It can and must happen in this way here too so the child can be pulled with its feet down below. As you can see in the seventh figure, the child was soon pulled out by the right leg up to its belly. Then with your right hand you need to take hold of the child's left leg, which is pushed against its torso, and pull so it can break through right away. So you have an unnatural birth, as you saw before, and which should be well-known to you and to all midwives since it is nothing new. These seven copper engravings show this kind of hard turn for saving the mother. If it comes to the extreme that you cannot reach the child's two feet and bring them together, then you must seize the first one and try in every way possible to get it, even if the child is not always lying crowded in so tight. Often a midwife has sturdy hands and arms; in that case it is all the more difficult for her to get at a woman, as I just reported, when the children are crowded in tight. Her sturdy hand and arm crowd it still more. Then she has to help as she can, namely, with a foot, as this turn shows, when she is summoned too late, if the woman is to be saved. I wish I could in your and my neighbor's best interest present all the memorable and remarkable things I know and inform you about them. I would not hold back a thing. But I have ignored a lot of things. When it comes under my care again, I recall it again. That is how it went for me with this delivery as well. I have had a number of these kinds of deliveries and yet I had not thought it necessary to present them until I was reminded of it just now. For when the danger has passed, you forget it was so hard and difficult. It is lamentable that it is so easy to make a mistake with such difficult deliveries—out of the plain ignorance of the midwife and the woman in travail. But mark this well if you wish to fill your office properly and preserve your good conscience.

CHRISTINA: Do tell me! Was the error in examining her responsible for this woman's ensuing difficult and dangerous delivery?

JUSTINA: If you but consider these last seven engravings well, along with the one previously presented, you will upon careful reflection acquire a grounding in it by looking at it, along with the written instruction, and you will find that an examination by a practiced midwife early on can prevent a good deal. My first admonitory copper shows you how a child mostly does not shift to a wrong posture until during labor for the many reasons I have already explained. To wit, when the child's hand is right by the head and the head lies crooked. If you look at figure 2,[86] you will see there that I pinch or squeeze the child's fingers, and if you read in the lesson why I did it, then you will find it was done for the purpose of keeping the child in the right posture for birth. For I have tried it often and have thereby prevented such danger. This prevention can only happen as long as the waters are still intact. For as soon as the waters break, the hand or the elbow enters the birth passage along with them and the head gives way and is bent backwards or to one side on account of the force of the throes when the waters have broken. Thus that sort of bad delivery must ensue; it cannot be otherwise.

CHRISTINA: Do you think it went that way too with this birth? The midwife cannot help not being summoned until after the waters have broken. If it is not possible to avert it afterwards, then the danger was already present when the midwife arrived. What can she do about your accusing her of this mistake?

JUSTINA: I am quite certain that this child lay as indicated in figure 2 when the waters were not yet broken, only that the elbow was in front of the head, as here the hand is, because I found the arm stuck in so far by the elbow (as the first illustration of the last turn indicates) that I could not possibly slip past the arm to reach the child's foot or leg and so I had to cut off the arm to save the woman. The head was also pushed all the way back, as can be seen in the same figure. Children like these usually lie right at first before the waters break and are still intact. I call it right even if it is somewhat crooked because it is still easy to help it out when labor starts and before the waters break. Thus I believe when I get that sort of delivery, when the child's hand or its arm has come out into the birth passage and I must save the mother, that these children all lay in the right posture when labor started and that mistakes were made and things neglected as a result of the woman's delay [in summoning me—*trans.*] and the ignorant midwife. A midwife can

86. Here and in the following speech Siegemund refers to figure 2 in the second series of figures beginning on page 94, which she compares with the figures on p. 131.

of course not help it if she is summoned too late and the waters have already broken when she arrives; that has happened to me. But as soon as she arrives, it is nevertheless her duty to touch the woman so greater danger does not ensue if there is something wrong. If all is well, then so much the better for the midwife and the woman.

CHRISTINA: And if the midwife had no understanding of the matter and the woman also would not allow herself to be touched, what would you do then?

JUSTINA: It is bad enough if a midwife has no understanding of the matter, and for that reason you need to learn about it. But stiff-necked and headstrong women should also mark well what kind of misfortune they can bring upon themselves, and likewise midwives who just wriggle their way through, midwives like this one, since she stayed with the woman for three days without touching her and left it to chance, as the copper engravings indicate, for you cannot perceive or prevent danger from outside or by looking. Therefore do not let yourself be talked out of touching the woman. If you encounter women who will not allow it, show them the danger they have to bear in mind here. If they will not, then let it be at their own risk. Your conscience is clear, and you can even abandon the woman. No blame can be apportioned to you, neither by God nor by the world. For he who will not listen to advice cannot be helped. But if they do not permit you to leave, then they have to permit themselves to be aided and they have to choose one or the other, inasmuch as these women really are in trouble. I have had to deal with women like that whom I had to force with stern words, and for these women themselves it did indeed end happily and they survived and retained their health. However, several of the children were stillborn as a result. Nevertheless, the women thanked me when they were saved. Once, however, I assisted a foolish and ungrateful woman in a hard and difficult delivery and undeservedly got a tarnished reputation as my reward. But what can you do with foolish and ungrateful people, since it is said of them, ingratitude is the world's reward. There is not a single office and profession in the world where things do not work the same way, nowhere where there is not a scabbed sheep to be found among them,[87] where you do not meet with ingratitude as your hard-earned reward and have to accept it.

CHRISTINA: Many women will be horrified when they read this book and properly contemplate the copper engravings showing the turning of the child, where you set to work in the belly with your entire arm as well as with

87. This appears to be a modified version of the saying "one scabbed sheep spoils the whole flock."

a rod with a ribbon wedged into it, as figure 17 indicates. It is hardly possible to believe that it can happen. Plenty of faultfinders will doubtlessly determine that it is impossible and cannot happen.

JUSTINA: I must confess that I, too, would be horrified if I had not myself practiced it with success, not just once, but many times, when I had to provide assistance. Thus I cannot think ill of a woman if she is horrified by it. And I wish that all ignorant midwives were horrified, not so they were afraid to learn it, but so they learned to prevent it, because preventing it is easier for stupid midwives to manage than that sort of difficult turn. But it is true that it is possible for a proper and experienced midwife to do it and that it could happen, and I could attest to it with the example of many women, if necessary, where I tried and used it with great profit and gratitude in such emergencies. Take a look at the testimony Frau Maria Lorentz gave in court, which is the seventh one, wherein such touching and intervention comes up, where I inserted my arm up to the elbow in order to help her and then demanded a smooth dowel, along with a ribbon.[88] You and those who read it will hear and see it is possible for the woman in travail to endure it since she is thereby helped and yet knows not what I did with it. So you can well imagine that it must still be bearable. For these women testify to that effect and attest to the fact that it was as I indicate in figure 17: the rod was inserted between the child and the midwife's arm. The woman did not really feel anything except pressure, which is not dangerous if only it is applied or imposed gradually. So let the faultfinders judge this as they will, as long as we can help and save our neighbor when Necessity demands it. Various people have persecuted and slandered me on account of this turn and indeed even said they would rather die than suffer me to treat them in that manner. I will not say what else they had the audacity to say. And it came to pass that they thanked God I saved them in exactly that manner; for this reason it is said, Do not be deceived, God is not mocked.[89] It is a mercy from God for me and for these women when they are saved and remain live and healthy. Not until then do they recognize what this presumptuousness is and what God can do through His grace and omnipotence with the ordinary means He provides.

CHRISTINA: The eighth testimony, Frau Barbara Vogt's, makes you think; her delivery was very dangerous. Explain to me at length how it went!

JUSTINA: Take a good look at the testimony she gave in court and think about it, for it explains pretty much everything about my interven-

88. See chapter 7.

89. Gal 6:7.

tion.[90] Take a good look at the lesson I gave you as well, where I showed you in detail how it actually went. You must have forgotten it. Look back on pages 86 and 87; you will find it there.

CHRISTINA: The ninth testimony is from Frau Susanna Jacob, née Ritter.[91] This woman enjoyed a happy outcome. But what caused her to be so long in distress? Did the midwife make a mistake by urging her to push too soon? She did say you have to await the appointed hour; so she must not have pushed her.

JUSTINA: From this you see the midwife had no grounding in how to touch her, and it therefore ought to be a lesson to you to learn to understand properly the copious information I have given you about how to touch a woman judiciously and thoughtfully. For you see in this testimony that a lot—indeed life and limb—is at risk for the mother and child should the midwife have no understanding of touching, whereas an experienced midwife can and must be guided by it in all circumstances, if she is properly to manage the woman giving birth. This midwife was of the opinion that the appointed hour of birth had not yet arrived. Here you see the error and its opposite. For when I touched her and helped out the child that was lying crooked, a happy birth ensued within two hours, and according to her testimony, she was in her fifth day of labor. Is that not an example of the midwife's ignorance, inasmuch as she recognized it as full labor already on the first day and on the fifth day she said, you must await the hour, however long it be. It is true that where there is no deficiency, you must await the hour, but in that case there are also no proper labor pains present. By contrast it is also true that you can miss the appointed time as a result of ignorance. Thus you must learn to note the nature of the fault through judicious touching if you are to exercise your office properly before God and the world.

CHRISTINA: When I have that kind of delivery, should I not give her something to make her push? If this midwife had pushed her, this woman might have delivered sooner.

JUSTINA: I clearly see you have not yet grasped my view. Did I say then that I administered something to make her push? I said I found the child lying with its head crooked and diverted it when I touched her and that the throes followed on their own. For usually if the pains are lacking in a birth where everything is in good order for delivery, then the child is obstructed in one manner or another. So it is necessary for you to guide the child rather than to push it. Pushing a child that is fixed would put it in danger of losing

90. See chapter 7.

91. See chapter 7.

life and limb. You see thereby what benefit judicious intervention can bring and how foolish intervention can be injurious.

CHRISTINA: I have to remind you of something else here, on account of the fifth testimony, which attests to the necessity of breaking the waters, where—so I see from the testimony—you so early on, indeed, several weeks before the delivery, suspected and announced an unnatural birth. And it came to pass, so I see, as you predicted. How can you know this so far in advance? Many of your enemies and those who do not wish you well can make up fantastic stories and evil gossip out of that. So recount how you came to suspect it and could know it.

JUSTINA: This is the reason why I suspected a difficult birth so many weeks in advance and why it turned out that way too. When, after the midpoint of pregnancy, frequent touching has not revealed children to have turned even once so they lie right for birth, unnatural births have always followed. It was like that with this woman. Since she had always had a bad time with miscarriages or, as they say in Silesia, she often teemed unhappily and would have the child before it was time, I was asked various times to give a report and always found that the child lay wrong, whereas after the midpoint children ordinarily tend now and then to lie turned to the right posture. The child is still tiny and thus has all the more room to turn over and over, which thing I have perceived with many woman: soon after the midpoint I have frequently found children lying right, especially when the women complained of pain and pressure. In contrast, however, when I touched them again the next day to see how things were when they were not having so much pain, I found the children lying higher up in the belly, sometimes in the right posture, sometimes in the wrong one, and this lying right and wrong so early on gave me great cause for reflection. I will express no opinion as to whether or not it is true that after the midpoint children usually lie in the womb with their heads down below, because soon after the formation of the fetus it is the largest part to be seen and up to the midpoint is usually larger than the child's entire body, and thus on account of its roundness and size, it drops down below the torso like a stone. As soon as children are completely formed and lie still, they usually lie in the right posture. When, however, they move, be it ever so little, they can turn downward and upward, however it goes. I have seen changes like this in most bellies, often two and three months and longer before natural delivery. Thereafter, however, when children have attained their proper size and the woman's belly is filled up with the child, then they are prohibited from turning upwards or downwards, and the last two months they always remain in the right posture if a natural birth is to ensue. In the case of those who can turn up to the hour of birth for various reasons, which I have already indicated to you, the midwife has to be careful during the delivery if an unhappy and unnatural birth is not to follow. Thus I think most unnatural births occur because in such bellies the child can turn over and over until the hour of birth. If the midwife does not realize this in plenty of time—and many women will not suffer the midwife to touch them early on until they pay the price of their experience with injury—an unnatural birth can easily result, whereas it could be prevented by breaking the waters at the start of the delivery, which is, as I have already explained, necessary when children lie right and the midwife knows they can turn in the woman's belly and thus cannot be relied on to stay put. There are also bellies, or rather children, as I should say, that I have not found in the right posture for birth during the entire pregnancy when I touched them. This occurs very rarely, and I can scarcely recall but a few cases like that. This woman, for whom I foretold an unnatural birth, was a case of that same sort. For I had observed previously that when children are not found lying in the right posture even once after the midpoint, an unnatural birth always follows. And precisely for this reason I thought this unnatural birth probable.

CHRISTINA: I need to remind you of something else here. Why do you not make use of the so-called speculum in order to screw open women to see how the child lies. I have heard so much talk about this, and you do not even mention it. What are your thoughts on it? I would think it would be better for you than using the crotchet; you will not hurt or hook the child with it, as happens with the crotchet.

JUSTINA: This is a fine indication of your judgment. You warn me and advise against the crotchet and you advise me to use the speculum to screw open the woman, when being screwed open causes women more pain and provides less assistance than the crotchet. If I see the child lying there, I certainly cannot help it by looking at it; I have to use another instrument for this, and so it is double the torture. And if an inexperienced midwife sees the child lying there, but does not know what she can and should do about it, how will looking at it help her? The speculum provides no other assistance than helping you to see what part of the child lies before or in the birth passage. You have to know that because with God's special help and blessing I got as far as I did in my apprenticeship with the poor peasant women and because I always had as happy an outcome as possible in many difficult cases in many ways, as I have shown you sufficiently, I never thought of a speculum nor did I know what I should or could do with it during the time of my apprenticeship. When I was farther along, namely, when I associated with people who could give me information about what to do with it, I no longer needed this aid, or rather, this double torture. As a result of a great deal of practice, along with meticulous examination, my hand and fingers had come to practice natural feeling or recognition so I could feel and distinguish—as if I saw it with my eyes—how the child lay. I thus consider my using the speculum an unnecessary torture. I do not reject using those kinds of instruments or the speculum in certain cases, as with all kinds of ulcers in the womb or the sheath and other injuries that cannot be so easily discerned or examined by touching. But it must cause a great deal of pain. I tried it once on account of an ulcer a woman had in the so-called neck of the womb or sheath. But pity kept me from undertaking anything I could have imagined doing with it. Yet Our Dear Lord saved the woman. For this reason, to this very hour I am afraid of the so-called speculum. To be sure I believe that when pregnant women are in their travail, it is easier to screw them open, because the fetus forces the widening of the birth passage with the pains and the sheath or neck itself has no such violent pains. But it is not necessary for me and my art; my gentle hand can lie where the hard iron lies, and it does not cause such pain. And I also do not see how, in turning the child as I do, my arm and hand could guide the child with the hard iron in the way. I think the people who need this speculum are the ones who remove children from women in pieces, using instruments. You need greater knowledge for that—and it is certainly also more dangerous—than you do for turning the child as I do, which Our Dear Lord deigned to have me learn among the poor peasant women. In all my days I have never had to cut a child to pieces in the womb. But when an arm presented, I have at various times had to cut it off, because the child was otherwise in too tight to be turned. Because in such protracted or delayed deliveries it is not possible to get the arm back inside, the arm gets in the way of turning so I cannot guide or turn round the dead child. However, as soon as the arm has been removed, it is easier to turn the child. The child's arm causes the child to get stuck, while my own arm hems in the child's arm in the birth passage so it is not possible to get it back inside when the birth is dry and protracted like that. The dead children are so firmly pushed forward and wedged in that I can hardly slide in my hand to guide the child. Here you should note on the subject of removing the arm from the dead child that it is often highly necessary to remove it for turning because you can turn the child more easily then, especially when it is lying wedged in very tight and has no more wetness around it. But it is not good to remove the arm until you know how the child actually lies. The arm that is wedged into the birth passage holds the child fast during examination so it cannot flinch, and then it is hard to feel how the child lies. If it is not lying crowded in too tight and still has some moisture around it, it will suffer itself to be pushed by my arm and intervention as if it were giving way. If it is still lying loose with plenty of room, it is not at all necessary to take off the arm. For, as I said before, the arm that is wedged in holds the child fast so I can reach the feet and put a loop around them, as can be seen in the engravings of turning the child where the figures demonstrate how the child can be turned without removing the arm. But the child must still have some moisture and slipperiness about it; otherwise it will not work. And it is precisely for this reason that the sooner the child is turned, the better and easier it is for the midwife and the woman in travail. Take note of this and you will manage it.

CHRISTINA: After the lesson you have just given me, I would think that if a child were not wedged in so tight and still had some moisture around it so it could retreat with the help of your arm and your touch, as you have just shown me, it would be better for you to get the child's arm back inside and guide its head toward a natural birth—you showed me previously how you can and should guide it in those engravings with the child lying with its head crooked. Then the child could in fact be born naturally and you would not find it necessary to undertake the cruel amputation of the entire arm and even less so to insert a rod and put a loop round the foot. That would be much easier for both you and the woman in travail.

JUSTINA: It would be desirable for you to be able to do this with dead children as you do with living ones. But it does not work that way, for when I guide a living child, it wriggles and thrashes in accordance with my guidance. A dead child cannot do that. So it would not help me if I were able to get the arm back inside, and it also would not be possible without a lot of pain. What good would that have done? Consider it rightly: the child is dead, and its head has been forced back above its body from the violent throes because its arm entered the birth passage; the child had to die as a result, and its neck was broken by the force of the throes and the wrong posture. This broken neck cannot be guided as you would with a live child, and the child also lies in a painful belly and is really wedged in tight, what with its waters already broken and for the most part run off, so you and I are prohibited from guiding it to birth by its head. The giving way I have described for you is only a giving way, but turning a child by the head and giving way are two different things. Turning a child by the feet is a completely different undertaking from doing it by the head, especially when the children are already dead. Legs are long, thin, and supple, and it is possible to guide them on account of the ankles. You cannot do that with the head, not only on account of causing the woman unbearable pain, but also on account of the size of the head in a belly that is tightly contracted. I refer you here to the testimony of Frau Thym as to how I fared with her when I attempted that sort of turn with a live child when the waters were not yet broken and where turning the child by the head worked and what dangers ensued on account of the child's lying too much in a ball as it had curled up round and crooked.[92] The midwife cannot prevent the child from curling up like this. Although it does not always happen, when it does, turning the child by the head, be it quick or dead, is always more difficult and dangerous than turning it by the feet, unless the child lay curled up with its body up above its head, thus lying only with its head crooked beneath it. In that case you can turn a head that is lying crooked toward the birth passage. But if a child lies completely wrong, you will probably have to leave off trying to turn it by its head.

CHRISTINA: Allow me to ask something else for it has just occurred to me again. You said earlier that you never had to cut up a child in its mother's belly, except you had at various times removed an arm that had presented in order to be able to turn the child. So I certainly see you do not think removing it is difficult insofar as the arm is outside the womb and you can see to remove it. But does not the hook you stick in the child's head while it is still in the womb frequently break the skull to pieces—as you yourself told me in the lesson you gave me?

JUSTINA: It is one thing to break the child's head to pieces with the crotchet when the child has already passed through the inner mouth and is lying squarely in the birth passage and when the top of the head can already be seen there, but the child is dead and cannot possibly be born on account of the mother's strength having been sapped; it is another to cut the child to pieces: its limbs are cut off one by one with the instruments used for that while still in the womb, way back in the belly. Because I do not know how to do that kind of amputation and because I know how to turn the child and have no need of it, I did not want to make mention of it to you. My thoughts are these: even if by screwing open the belly an inexperienced midwife were to see the limb of the child that was presenting, but she knew not how to guide it, seeing would not help her. Indeed, if you could see the entire child lying in the belly by screwing it open and could discern how the child was lying and were there space enough to guide it, that would be good and desirable. But you can only see the child's limb, and it is certainly harder to discern when it is lying deep inside because it is dark. It is, however, impossible to know how the child lies by looking with the aid of screwing open the belly until the limb, be it the arm or a leg or whatever, is amputated. An inexperienced midwife must and will refrain from this sort of amputation, and

92. See chapter 7.

I consider it to be more difficult and dangerous than the turns I have shown you. Nevertheless, I leave each person to practice the art in his own way, as Our Dear Lord has made it manifest to him, and as he best knows how to offer assistance. May God Almighty grant all of us His blessing.

CHAPTER 6

Of the afterbirth, whether it can and does lie in front of the child and whether it is attached or can always be gotten out

CHRISTINA: Dear sister, tell me what the afterbirth is like and whether a woman can be endangered by it.

JUSTINA: It is necessary for you to note that it is a bad and dangerous delivery for mother and child when the afterbirth comes into the birth passage with its thick piece of flesh or sponge (as I call it) before the child does.[93] I do not know why or how this comes to pass, whether it comes from the mother jerking or slipping or whether it is possible for the child to tug on it. I have experienced various such deliveries. They are very dangerous for mother and child because prodigious flooding[94] can occur. I have been summoned to a number of them when the children were already dead from the flooding and the mothers themselves did not have much vitality left. This condition is seldom understood to be labor because it is not especially accompanied by pains as the flooding weakens mother and child. An inexperienced midwife cannot accomplish anything because the afterbirth feels like the rest of a woman's flesh or the neck of the womb. She thinks it is not yet time for the birth because she cannot feel anything of the child, even though it lies right behind it. The sponge cannot give way, and the bleeding cannot stop until the woman is dead. As far as I know, they cannot be helped in any other way than for me to pierce the thick flesh of the afterbirth with a thin hook or wire or hairpin. (You have to take care that the hook or needle does not penetrate as far as the child.) The waters flow then as they usually do when they break. As soon as it gets air, the throes set in and the bleeding stops. Then I have used my fingers to part the sponge. Thereafter a happy birth ensued for mother and child if I got to them when they still had sufficient energy.

CHRISTINA: Is it always the fault of the afterbirth when women in tra-

93. By the thick piece of flesh, she means the placenta. She here describes the condition of placenta previa, that is, the placenta is implanted centrally so as to cover the internal cervical os.

94. Hemorrhaging.

vail flood? This often occurs a few days or a few hours before the birth, before there are any pains. What is to blame for it? Can it be helped with medicaments?

JUSTINA: This or similar flooding close to or during delivery can more readily be stopped with medicaments than can the other kind. I had this kind of flooding in the case of a gentlewoman, when the afterbirth had not come out first. She was to be confined with her first child. She had frequent flooding before there were any pains, and it was so prodigious that within three or four hours the woman was at death's door. God alone, not I, knows what the cause of this flooding was. I could feel no sign of life in the child. The woman began having hallucinations. Her hands, feet, and head had become icy with sweat so there was little or no hope of saving the life of mother and child.[95] One could hardly feel the pulse and sometimes for a long while not at all. Medicaments no longer had any effect, and so, as I said before, there was no more hope in mortal eyes. Nevertheless I suggested the following means to the venerable doctor. I wanted to take a hairpin and pierce the skin that retained the child's waters inasmuch as during birth flooding tends to stop. Whenever I had pierced the afterbirth so the waters ran out, the fetus, dead or alive, had pushed forward for delivery. The venerable doctors decided I should do it. As soon it was done, the waters ran off as usual, and the child, because it lay right for birth, pushed into the birth passage with its head, and the flooding stopped. Thereafter there were some pains and I could expect a happy birth soon. But because the woman had been sapped of her strength and the child was dead, as you can imagine with such powerful flooding, the pains had not the force to continue and ceased again. After the flooding had also ceased, the woman lay for several hours in a deep sleep, nearly unconscious, until along toward dawn she recovered somewhat and came to her senses. Pains set in once again, but because the woman was weakened and the child was large and dead and the woman was giving birth for the first time, it was impossible to deliver her, even though the venerable doctors and I vigorously extended her every assistance with fortifying and stimulating physics. Then the venerable doctors determined it would be beneficial for me to hook a crotchet in the dead child's head and draw out the child in any way and as best I could and thereby save the mother if possible, which thing I performed successfully without in any way injuring the woman's health, though she lay in bed for nearly half a year before she recovered. But to date (God be thanked) she is hale and hearty. So

95. The condition that Siegemund describes here is probably eclampsia, that is, pregnancy-induced hypertension accompanied by grand mal seizures.

you can see it is not advisable to repudiate the use of instruments across the board; as long as they are well employed at the appropriate time—gently, cautiously, good, and carefully—then Our Dear Lord will bless such work and aid as well. Knowledge is not to be repudiated. I have attended many difficult births and cases among poor peasants who begged me, for the love of God, to help them. So I attempted the aforementioned measures. And because God blessed them with his grace and aid, I tried them with other women and other deliveries as well, as outlined here, and found them to work well, for which I heartily thank divine blessing and aid. I have studiously refrained from talking about natural births, which are besides well-known, and we have reason to thank God that such difficult cases are not common. I had probably had thirty cases and more of natural births before I met up with one such case. Indeed, something like that happens hardly once in a hundred cases. The reason I have so often been summoned in such births is that I have a reputation for being able to provide assistance. So I have been fetched a number of miles to attend such cases if it has been at all possible to get hold of me. Thus I have become more and more famous up to the present day.

CHRISTINA: How is it that with some women the afterbirth is slow in coming after the birth and with others it does not emerge at all until they have died of it. What do you think the cause of it is?

JUSTINA: I will gladly share my knowledge of it with you for I have been summoned and fetched many times when the woman's life was endangered because the afterbirth remained inside after the birth. Over many years I have found no other cause but the falling of the inward mouth, which could then be aided with two fingers. You just have to do it carefully and gently, namely, if the afterbirth will not come out as usual, I see to it that the fallen belly is lifted up and held and then I take the navel string in my left hand so I can follow it with two fingers of my right hand up to the mouth. Then I push it up as it was during the birth and thus the afterbirth gets air and drops. It is necessary to exhort the woman to cough. But it is more necessary for one than for the other. You soon know by feeling the navel string and the afterbirth whether or not it is necessary. If it is necessary and good, then the afterbirth and navel string will descend. If they recede from this coughing, then do not have her cough anymore. Otherwise the navel string will tear off as a result of being shaken like this and it will be very dangerous. To my knowledge there is no better advice than not to let the inward mouth descend and to hold onto the navel string, but not too firmly so it will not tear off. It is also good to lay something warm on the belly, warm wine or warm, buttered beer. But you must take pains to keep the belly warm, for the warmth helps a great deal. I have done this faithfully from my earliest beginnings and found it beneficial. Now I have not found it necessary for a long time because I can search for the afterbirth myself when necessary. I do not know how to teach this if common sense itself does not teach you because it is difficult to recognize the afterbirth in the soft belly. However, the navel string will lead you to it so you will find it upon careful reflection. But you have to proceed cautiously so as not to tear it to pieces. It can happen quickly, and then grave danger sets in. It is not good to be too hasty; it is also not good to be too slow. So you have to reflect carefully on what you are doing when you employ this method. After a lot of practice you will learn as I did that you have cause to thank God on High. I gladly confess that I worry more about the afterbirth and getting it out than about turning the child in any way. No matter how the child lies, I have been able to offer assistance when the case has come under my care. May God continue to grant His help! But with the afterbirth I have had two unhappy cases as follows.

I had delivered over six hundred children before I treated a case where the afterbirth was attached. Furthermore, in all these deliveries, I had already been fetched over one hundred times to assist in perilous cases and yet had never had the misfortune to be unable to save the mother. Thus I became so brave and bold as to believe there was not an afterbirth to be found that was unnaturally attached. For in all the cases to which I had been summoned and that I had treated, I merely found the inward mouth had fallen. So I disputed it vigorously when I heard people speaking of attached afterbirths, until one day God visited it upon me. It happened in Liegnitz with a brewer's wife who had given birth to her twelfth child. She had had an easy delivery so I scarcely attended to her for half an hour. However, when I set about cutting the navel string, it was so short I hardly knew how to cut it. In the end I had to cut it as best I could. I did not want to let the navel string slip back into the womb so I tied a ribbon to it so as not to lose it. When I had handed off the child, I set about, as usual, trying to get out the afterbirth. I had the woman raise her belly, as I have already explained, took hold of the navel string, searched for the mouth, and exhorted the woman to cough. When, however, her belly shook, the navel string receded with such force into the belly that I was mightily frightened and knew not what it could mean. But since I still had the navel string thanks to the ribbon I had tied to it and since I knew how to search for the afterbirth without causing injury, I followed the navel string to the afterbirth and found it in the fundament of the womb attached in such a manner as if a clump of blood had coagulated firmly on something. Because I had never before felt anything like that with the afterbirth, I was more frightened than before. I withdrew my hand to ponder what should be done here but could not make a decision on my own, insofar as I was afraid if I tried to get out the afterbirth, be it by manual intervention or by giving her strong medicaments, a great flooding might ensue. I did not want to take on the responsibility for this on my own. So the venerable town doctor was called in. He found it advisable not to act hastily inasmuch as the afterbirth often remains inside for several days. Perhaps it would detach itself gradually. Even as we were conferring in the other room, a screaming arose that the woman was dying. We hurried into the lying-in room and found the woman in dire extremity. When I started rubbing and cooling her down, she revived and did not know what was the matter except she had had a pain and thereafter felt good. But when I set about examining her to see whether the pain with this violent sickness had perhaps separated the afterbirth, I found her lying unnaturally in a pool of blood. The navel string was as short and as impossible to get hold of as before. It was not long before there was a new pain, along with violent flooding and dire extremity. The doctor attended her as well and prescribed something for the flooding as best he could. It would not be stopped until she passed away in the fifth hour. This was the first misfortune I had had in all my time. Thereafter I delivered more than two hundred children and did not encounter such a condition. However, I was not past it, so for the second time God taught me a thing or two, as it were, with the afterbirth with Frau Lorentz, once again in Liegnitz. But it was in this manner: the afterbirth would not come out so I searched for it as usual, finding nothing in the womb but the navel string, along with the skin that encloses the waters, both of which were attached to the womb and completely dried out. I could not feel anything of the cake, which is usually the largest part of the afterbirth. I could think of nothing I could do or of any solution. I had the doctor summoned once again. He did not know what to do either. In the sixth hour after the delivery, alternating weakness and nausea set in. This continued until she had convulsion fits too. Near the end, as her grave illness persisted, blood spurted from the neck[96] until she passed away. To this very hour I have lived in fear, though since that time God has graciously preserved me from such occurrences and I have cause, and every midwife along with me, to entreat

96. Vagina. The language that Siegemund uses is ambiguous, for in the seventeenth-century gynecological context the German word *Hals* can mean either throat or vagina. Given the failure to remove the placenta, bleeding from the vagina is far more likely, although in early modern conceptions of the female body, what cannot emerge from the vagina will come out of the throat, and it is therefore possible that vomiting could have been perceived as bleeding from the throat.

Our Dear Lord for His blessing and divine aid. Otherwise I can say with good conscience that I have never encountered a child in the womb that could not be aided by turning so at least the mother was delivered of the child and saved. I thank God from the bottom of my heart and entreat Him to continue to stand by me, to protect and bless me. But this condition of the attached afterbirth was more difficult and dangerous for me to treat—and it was impossible to save the mother—than all the other deliveries and turning of children. May God preserve me and all people from such peril.

I must nevertheless once again recall attached afterbirths inasmuch as I recently encountered such a case. Since the aforementioned women had both lost their lives when the afterbirth was attached, I resolved this time to try peeling it off, if it would work, since by all appearances this woman would die anyway. So I took hold of the navel string firmly with my left hand, but not so firmly that it could tear, just firm enough for the afterbirth to be slightly lifted up by the navel string attached to it. Hereafter I introduced the customary fingers of my right hand alongside the navel string, and as far in as my fingers could reach, I gently scraped all around between the afterbirth and the womb. Then I employed my second finger, enlarging the hole and scraping the afterbirth thoroughly until it came off. Thus this woman was saved. She lay sick for a long time before she recovered. I do not know the reason for her lying sick for so long. However, I believe it was because this scraping was a very delicate matter and because the afterbirth could not be scraped off completely; indeed, many stubs or bristles were left behind. These pulled out of the thick cake during the scraping and remained fixed in the womb. They were certainly not large or long but rather short and sharp. They felt like a rasp. I am convinced that these little filaments, which remained behind, led to a putrefaction that the woman could only slowly overcome. Nevertheless, she survived and was thereafter hale and hearty. Whether this could be repeated is in God's hands. So I think in this circumstance and with that kind of assistance there is always grave danger even if things go the best they could. No medicament seems to work well; it cannot, since the afterbirth is attached. Stimulants do not detach it either: flooding ensues, and still it does not come forth. Even without forcing it, flooding readily ensues and cannot be stopped until the woman dies, which I have seen with my own eyes. So, given such conditions, it is, to my knowledge, dangerous in every respect.

CHRISTINA: Do tell me how it is with twins, whether they always have only one afterbirth or whether each child has its own afterbirth and whether, if they have only one afterbirth between them, the waters break twice. And what happens if each child has its own afterbirth? Does the first afterbirth come forth soon after the first child or does it stay inside until both children have been delivered. Is there something I need to know here?

JUSTINA: Twins, to be sure, do have one afterbirth and only one membrane or skin; but each has its own waters, for the middle skin of the other child retains the waters. When the first child is born, the entire water bladder of the other child follows. When the birth of the first one commences, it has to break, be there a little or a lot of water with the children. Of course it is possible for two children to lie in one water bladder and one afterbirth with no skin between them, but this is rare. I have assisted in the birthing of many twins and can say the waters break only once and one child comes forth soon after the other. Many other twins even have an afterbirth each. In this case the waters break twice. If the birth is managed carefully and properly, when the first child is born, the waters of the second one expand, like the first, until they break.

CHRISTINA: You say the second water bladder follows soon after the first child when the child is delivered properly and managed carefully. What should I understand by "delivered properly" and "managed carefully?" Tell me what this involves.

JUSTINA: Delivering it properly involves a reasoned and thorough examination. This will tell you whether an additional child is present or not and how it lies. An external examination will of course also tell you whether an additional child is present, but not how it lies, which is of course what is most necessary to know if the child is to be delivered properly. Before you retrieve the afterbirth of the first child, you have to know how the other one lies since the retrieval of the afterbirth can cause the child to turn even though it was lying right for birth. The navel string attached to the afterbirth is very long and so the afterbirth will remain inside because, as often happens, during birth a long navel string does not pull on the afterbirth so it follows the child. Thus the afterbirth remains inside because of the long navel string. Then the second child pushes forward for delivery soon after the first one and a happy birth soon occurs if it lies right. If the navel string is too short, as mentioned before, the afterbirth follows soon after the firstborn when the labor pains come. Thus the afterbirth that is at hand and in front of the other child is drawn out first. The other child cannot otherwise be delivered until the afterbirth is out of the way. A sound examination will tell you whether the afterbirth is lying in front of the other child or the other child in front of the afterbirth. But the touch can mislead you in such circumstances, for if the children have but one afterbirth, the cake is very large and extends down to the inward mouth so you can reach it yourself when you touch the woman if the first child has already been delivered. You can be deceived if you think it is a separate afterbirth because it appears to be lying in front of the other child. Touch the woman with considered reflection and you will find it lies to one side against the rectum and that the other child is easy to reach farther up against the woman's bladder. In such circumstances you need to be careful, for if you pull on the afterbirth a little by the navel string, as is customary, you will displace the other child, which was lying right, and the afterbirth will settle down in front of it, causing the delivery to last two or more days longer. The child cannot be born, neither in a right posture nor in a wrong one, until it can get the afterbirth behind it or to one side through vigorous motion.[97] At that point it will be born, and the birth will take place in whatever posture the child lies. This prolonged and dangerous labor can be averted from the start if you, as I said before, find the child above the woman's bladder and aid it. The birth will follow soon as well, if the child is in position. If it, however, is not in the right posture, you must help it into the right one as well as you possibly can. You would similarly jeopardize the delivery if there were two afterbirths, as mentioned above, where the navel string is very long, so the first afterbirth neither yields to the child nor follows it during delivery. In that case the second child will come right after the first one and the first afterbirth will have to remain inside until the other child is also born. If a midwife is not aware of this and tries to get the first afterbirth, as is customary, she will either tear the navel string off the afterbirth, or if the navel string is so sturdy that the afterbirth comes out after it, she can easily displace a child that lies right so an unnatural birth follows. This is why you must gain knowledge of the second child after the birth of the first one before you try to get out the afterbirth; if you do this, you will not go wrong. And you have to do the same if still more infants are present. With triplets I have found two and three afterbirths: two shared a single afterbirth with only one skin between them, as is usual, and the third had a separate afterbirth. It is also possible for each child to have its own afterbirth, so you have to be careful, as I said before, when you try to get out the afterbirth because you can jeopardize the delivery. If you know not how to touch the woman, you will have to let it happen as it will. But if you have a foundation in and a good understanding of touching, then you can avoid all the aforementioned hazards.

CHRISTINA: It is astonishing that so many ignorant midwives manage amidst so many dangers and that we do not hear about more unhappy births. If I rightly ponder all the complications and bad deliveries you have explained and described, I have to think no birth can take place without one

97. Fetal motility was thought to aid in birth.

or the other of these many dangers. I could easily assume the mind-set of those who say of you, "That Justina has quite the silver tongue; she can put one over on people." I say this to you only to inform you and hope you will not take it amiss. Indeed, they say more than that, but I do not want to hurt you by repeating these things. Since, God be praised, you are for the most part lucky in your profession, I can easily believe that people of ill will are envious and do not wish you well. But as the saying goes, fear God, do right, and fear no one.

JUSTINA: It is not my intention to show you the good and happy births, for these take place every day (God be praised in Eternity!) on their own without the art and knowledge of the midwife. If that were not so, then it would be impossible for so many ignorant midwives to manage, and few women would survive and retain their health. And God is to be prodigiously thanked that births go well nearly every day and on the whole. We can see God's grace therein and we cannot thank Him enough for it. But God also frequently allows unhappy births to come upon women and yet can aid them. So He has arranged for natural, or rather proper, means to do that. So I wanted to make an effort to show you these unnatural births in order to teach you. For the lesson and the aid are highly necessary and give us reason to thank Our Dear Lord. And it is truly said of the birth of humankind, great works are the works of the Lord; whosoever esteems them has pleasure therein.[98] Indeed, Our Dear Lord is kind and merciful and shows and gives us means and ways to help when He punishes and burdens us. For if births always went equally well, people would think they did it on their own and would not recognize the almighty power and mercy of the Lord. And it often comes to pass that women who have happy deliveries attribute it to their own precautions and help, and they say, "I do such and such, indeed, I take good care of myself, and when the time comes, I have no difficulty giving birth. It is those milksops who are afraid of the work." I knew a woman who had quite happily given birth to eight children who always said, "What a silly fuss women make when they give birth; I can give birth like nothing." But Our Dear Lord let her lie five days in utmost extremity with her ninth child, and because she was not aided by having the child turned—I was away at the time—in the end she died along with her child. All those arrogant children of the flesh need to consider this and not profane God's mercy. For it is not said without reason, do not be deceived, God is not mocked.[99] I think the same thing about these people who speak so ill and ignorantly of

98. Ps 111:2. Siegemund slightly paraphrases the second clause of the verse.

99. Gal 6:7.

me. They will pay heavily for it. I hope no one will find anything amiss in the lesson I have given; I have given it with good intention for the benefit of all humankind, as observed in my daily practice and as assiduously as possible, and published it at the demand of many pious women. May God on High bless this work and let it turn out to be to His honor and the good of my neighbor. Thus they will protect me against all slanderers who harbor unchristian and unjust opinions about me and say my aid is unnatural and thus satanic. I hope nothing that is unnatural or unchristian can be found in or inferred from my prodigious and difficult work. Nevertheless, I have to recall the proverb, Better envied than pitied. May God convert all sinners, support the just, and be merciful to us all! The fact that midwives for the most part can and do manage is not the result of their knowledge but of God's grace. For He is the creator and preserver of the race of man; thus He rules over childbirth without the aid of human reason. But since God gave humankind reason and charged each person with an office and a profession, the one this and the other that, each person must be mindful of his profession if he wishes to live a Christian life and to die redeemed, and especially midwives in that this work involves life and limb of mother and child as well as women's health. Thus it is lamentable that there are some midwives who do not think about what they are doing and know no more and wish to know no more than how to receive a child when it falls into their hands and how to cut the navel string. They do not concern themselves with anything more, even violently dispute the possibility that a midwife can do anything more, because it is hidden from them. Precisely this pernicious dispute and the many dangers menacing woman and child in the hour of birth moved me to take laborious pains to assemble this manual by the grace of God to inform and benefit everyone. It is true that Our Dear Lord aids us daily and can aid us and that He need not employ natural methods if He wishes to help us. Nevertheless, He gave humankind proper ways and means in all things that we should not repudiate. For Our Dear Lord frequently will have a pious woman get into dire straits in childbirth, perhaps for the sake of her salvation. Thus he gives us these natural means to help them visibly since our reason seldom recognizes the invisible means, and God is little thanked for these. So that is why He often lets us perceive the danger so we will recognize and accept natural aid with gratitude and not repudiate it as often happens, although even natural means are nothing without God's blessing. That is why it is said, Pray devoutly but hammer stoutly. Thus the Lord will bless you: Unless the Lord builds the house, those who build it labor in vain.[100]

100. Ps 127:1.

CHRISTINA: Tell me what you think about cutting the navel string? What is best? Leaving it long or short? Some midwives do not cut the navel string until they have removed the afterbirth from the woman. Is that right or wrong, and why do they do that?

JUSTINA: There are many views on cutting the navel string. Some say that if the navel string is left too short, the child will be short of breath. Others say that if it is left too long and a lot of blood is in it when she cuts it and if the midwife fails to push and stroke this blood away from the child or make it run out, it is very injurious to the child. Such children, it is said, are subject to ulcers as well as chicken pox and measles and many such rashes on account of the blood that becomes putrid and stinking in the navel string inasmuch as thick, fat navel strings often smell very bad for many days before they come off, which thing cannot be prevented. Some say too, the longer the navel string is when it is cut, the higher the child's voice. Thus there are all kinds of opinions in the world, and it is best in the case of things like that, which do no harm, to let everyone have his own will and thoughts. Who teaches the peasant midwives to cut the navel string? Often they cannot read a word and thus can get no information about it. I have come upon cases where they cut it off so short they could hardly tie it off; they were frightened and forgot they had to tie it off until after they had already cut it. I have also seen cases where the navel string is accidentally cut so short that there is little left of it to tie off, especially with these midwives who as a rule cut it off short anyway, for they say, "The navel string is too smelly. It is not good for the children to leave it too long." I myself have had the experience that in dire extremity I have unintentionally cut it off short. But because as a rule I leave it long, I have not had difficulty tying it off. I am bringing this up as a point of information so you will be careful and leave on a little more of the navel string rather than cut it off too short. You can always cut off a little more if you have left it a little too long, but you cannot put more back on.

I have also variously had the experience that children came forth before I could get to them and still it turned out well. But twice I have had the experience that the child came while the woman was standing and fell out of her onto the floor. Because no woman, indeed not a single person, was with them, the fall caused the child to be torn off the navel string. It was so short I had a lot of trouble tying it off. It is not good for it to be torn off like this, and yet these same children survived and became adults and exhibit no defect, none concerning the navel or with their breathing or with their lungs and voice. I have purposely paid attention to this. Here I would like to give such women something to think about: it may not always turn out like this if women deliver while they are walking or standing and are by themselves. So they should lie down as soon as they feel it coming on. Indeed, if they can go no further, it would be better if they lay down on the ground right where they are than if they remained standing. A child would probably fall to its death if God's kindness were not so great, and it could easily bleed to death as a result of the navel string tearing off deep within its belly. A woman in these circumstances can preserve herself from such peril if she heeds my advice to lie down. It is bad enough that such things can and do happen when a foolish or careless midwife is in attendance. I want to report as a warning on two examples of such unhappy births in the case of two women I knew well. When the one became ill with her first child and would not stay in any one place, neither in bed nor on the birthing chair, the midwife let her do as she pleased because she did not think the child would come under these circumstances, since with her limited knowledge she had not much of an idea of the situation. So they were all reassured and let the woman go. But before they knew it, a pain came, and the child fell on the floor. It did not die immediately, but rather a few hours later, so they managed to baptize it. It went almost worse for the other woman. This one had always had easy births and had a well-known, good midwife in attendance who had served her before in the births of six children. Because she had been accustomed to having hot boiling herbs placed beneath her as a fomentation she did it this time as well. She has scarcely placed the herbs beneath her or is about to do it, when a pain comes. The woman screams at the midwife, "Do something! The child is coming!" Even as she reaches for it, the child is already lying in the hot herbal bath. It is burned so badly that it dies within a few days. I am telling you this so you will know that even with quick and easy births you need to take precautions. And even the best midwife can be unlucky, if you expect too much of her, or Our Dear Lord withdraws the hand that blesses. Thus everything depends on God's blessing.

CHRISTINA: I gratefully accept your admonition. But give me a complete explanation of your view of cutting the navel string as I entreated you.

JUSTINA: My view of cutting the navel string is that it is best to do it in the middle, not too long and not too short. But you can do something more about having left it too long than having cut it too short. Navel strings that are too thick certainly smell bad. You cannot do anything about that unless there be a lot of blood in them, in which case you need to stroke back the blood before you cut it. If you cannot get the blood to go back, then knot the cord so when you have a better opportunity you can open it up again. Besides, thick navel strings need to be handled with care because they readily bleed, and children can easily die from not being tied off good and firm. Then you can loosen the knot, as often as you like and as is necessary, and tie it tighter. Thick navel strings settle two or three times to get air and they start bleeding again even when you think they are tied off good and firm. I had the frightening experience of a navel string that started bleeding again a full hour after the birth when the infant had already been swaddled. It would have bled to death before I realized it if it had not started moaning. When I attended to it, I saw it had gone completely white. When I took off the swaddling, it was swimming in blood. If it had taken me only a short time longer to become aware of this, the child would have died. Since that time I have watched over children assiduously during the first hour, and this is why I am warning you.

I have to admonish you about yet another thing, namely, that you be careful when tying off the navel string with cords that are twisted together too tight and fine or flax that is spun tight and fine because they tend to cut through thick, swollen navel strings. You think you have tied it tight and then it bleeds where it is tied. So if you see that the cord threatens to cut through the navel string or has cut through it, you need to tie it again behind the first tie so it is more secure. It is easy for this to happen with thick navel strings. That is why it is good to leave the navel string long so you can tie it again better and farther in. Otherwise I do not think there is much to say about cutting the navel string, be it short or long; besides, with all children the navel string rots off.

CHRISTINA: You still have not answered the question why some midwives do not cut the navel string until the afterbirth has been removed. What do you think is best, and why do they do that?

JUSTINA: I think it best to cut the navel string as soon as the child is born and to give the child over to be wrapped in warm swaddling clothes so it will not catch cold when it is removed from the mother's warmth. And it is also better for the midwife and the woman in travail, for you can cover up the woman better and protect her from the air than when the child is lying between her thighs or the midwife has it on her lap down below her.[101] It is likewise better for the child, especially if it is weak. For various reasons such a weak and half-dead child would often sooner die or be at death's door before the afterbirth can and does come out. You cannot refresh and warm a weak child on a cold lap as quickly as in a wooden trough and a bracing bath. Indeed, a weak infant put in a warm bath recovers more quickly than on a person's lap even if it is immediately rubbed with spirits or brandy. You can

101. Here one needs to imagine the midwife sitting on a stool in front of and below the parturient mother.

do this just as well in the bath as outside it, and it is much better with such half-dead children. I have often asked them why they have the child wait for the afterbirth, and they have answered that the afterbirth tends to come earlier on account of the movement of the child. I will certainly leave people to their own opinion, but if I may express my own thoughts, I must say I do not see how in the case of a weak child I can hold up the cutting of the navel string for the afterbirth. A weak child hardly moves or not at all. How then can it help the afterbirth come out if the movement of the child is supposed to do that? And often the navel string is so long that the child on my lap or between the woman's thighs would have to jump and dance before the navel string would move all the way into the mother's womb to get out the afterbirth. If it should be the case that the blood that occurs naturally in the navel string can do it as a result of the movement, then I fear it is too weak to do that if there is any complication or delay with the afterbirth. Even when the child is born hale and hearty with all its powers and the navel string is filled with blood, I have often had a lot of trouble getting out the afterbirth and have quite enough trouble using my hand—which is, thank God, rather sure of itself in the necessary touching and feeling—to get it out without injuring the woman, as I have already described to you in various cases. I will maintain silence on the question of what the blood in the navel string with weak and half-dead children can and should be like. Often in the case of such half-dead children there is not a single drop of blood to be found when the navel string is cut. I myself have had such cases of weak children where I did not find a single drop of blood in the navel string until the child recovered, one earlier, the other later, however they could. Not until the child had revived did blood flow into the navel string again. However, the blood can also flow into the navel string when the child is dead. But none will come when the child hovers between life and death. Thus the blood cannot, as some midwives believe, possibly help get the afterbirth out. For this reason I advise you and others to pay no heed to the opinion that says children are not to have their navel string cut until the afterbirth comes out unless it should come forth right after the child, which does often come to pass. For if it gets even a little stuck, then it will take a little time. Even if the child is hale and hearty, not to mention if it is weak and half dead, it will get cold fast because it has emerged from the warm womb, and this is highly injurious to the infant. The earlier the child can and does get to a warm place again, even if it is only in warm swaddling clothes, the better it is for the child. And it is precisely for this reason that many people have a good opinion of the bath because the child is kept at the same temperature until it recovers. This should not be repudiated for weak children, as I have found it a great help in those cases and believe it to be highly necessary. Stick with the notion that it is better to cut the navel string soon after the birth. You can of course feel for the afterbirth and tell whether it will come forth soon thereafter or not.

It has happened to me, and there are bellies, though not many (it would not be good either), where if you do not reach for the afterbirth the moment the child emerges from the womb, the inward mouth cramps up and closes so you cannot get the afterbirth out for several hours, even for several days. And if you but quickly introduce your hand into the inward mouth and immediately somewhat firmly restrain the navel string, insofar as it is possible, you have prevented it from happening, and the afterbirth will come forth right after you touch her if only you do not permit the mouth to close up. This cramping is merely an unusual pressure from the afterbirth. As soon as you make way for it through the inward mouth, it will immediately push forward and come forth. The child cannot accomplish this preventative measure through its movement. Thus the careful and intelligent assistance of the midwife is best, given God's mercy and His blessing, without which we have nothing. On account of this necessary intervention and such dangers, I purposely had the tenth and last testimony of Frau Maria Müller, née Ritter,[102] printed and included here to show you the truth. For precisely such a condition, that is, cramping in the inward mouth, created or rather caused the danger. So be informed by this that you must heed the afterbirth as soon as the child is born. Even if you cannot always get it out right away for all the reasons I have enumerated, you can nevertheless feel and prevent cramping in the inward mouth. If the child is weak, then have someone else cut the navel string right away. You cannot put this off until the afterbirth has come forth. If the child is hearty, then let it lie there and wait until you can cut the navel string, as long as this is not delayed for too long. If it is somewhat delayed, then have the navel string cut and the child taken away because it is just in the way, though such circumstances are somewhat unusual. Nevertheless, I have become very fearful of this situation inasmuch as it is hazardous and yet can be prevented, as you see from the testimony supplied. It was just such a circumstance and thereafter could be prevented from occurring a second time as a result of the instruction I gave the midwife about it and for which the woman thanked me from the bottom of her heart. So I hope to receive such thanks as well from you and imperiled women, if you and they heed my lesson and guidance.

102. The tenth testimony in chapter 7 in fact lists the maiden name of Maria Müller as Roth not Ritter. The maiden name of the previous witness, Susanna Jacob, is Ritter.

CHRISTINA: Dear God! Do tell me how it happens that many women always experience convulsion fits or the misfortune, as it is called, in childbed and why this continues as long as the pains do until they are delivered. And why is it other women cannot be delivered and mother and child remain together and must die together, but then the child emerges from her after they have died, of which occurrences there are many examples. How does that happen? Is there no means or intervention to prevent it? Know you of nothing of the sort? Tell me your thoughts on such situations.

JUSTINA: I think it generally happens to women who are naturally inclined to convulsion fits,[103] though it can happen to others as well, be it a case of an unnatural birth or a child that has gotten fixed as the result of the unusually violent throes common to such complications. When the birth is delayed as a result of the child lying wrong or getting fixed and when it cannot push forward and down into the birth passage, the pains shift upward and constrict the woman's chest so she cannot breathe at all. So then it is easy enough for convulsion fits to set in with women who are weak by nature. Many a woman may appear to be large and strong in body and nevertheless be quite weak. If a woman is by nature inclined to convulsion fits and has complications like this on top of it, then it is all the worse for her. And even if such complications do not ensue, and she has only the normal birth pains, fear and terror and pain generally come together. So these three things assault her all at once. Whenever the woman begins to feel the throes, she immediately begins to shiver and shake. Then the convulsion fits readily set in as well. Because I have observed this, I have exhorted and encouraged these women more than others and have been more careful when touching them than with others. For a midwife can sooner feel the woman's labor pains when they start or are about to start than can the laboring woman herself. So as soon as I felt the throes were about to start, I spoke to her in the following manner: "My dear child, be not afraid of the pains and be not afraid, be as tough and confident as you can and let not your courage flag and your hope fail. I assure you, with God's help it will go better than you think! Just hang on tight with both hands so you tremble not. It will soon pass. You will see that Our Dear Lord will soon help you. How quickly a pain passes! Who should let his courage flag so quickly, for God's help is at hand?" This encouragement along with skilled assistance has always greatly helped me right up to the final bearing pain when the child passes through. But it is not possible to prevent this when the woman has a natural tendency to it.

103. Siegemund probably speaks once again here about eclampsia, which is characterized by grand mal seizures.

But then the child is in such a condition that the skilled midwife can intervene and help it out. God be praised and thanked! I have never had a mishap with such complications, neither with mother nor child, when I had attended them from the start and when the children lay right. But whenever the child lay wrong, I turned it as soon as the waters broke on their own and helped it to delivery as well as a person can, and I spoke to the woman and encouraged her as already described. In this manner I managed a happy outcome. Nevertheless, not every child survived because some lay wrong, but nothing happened to the women. May God continue to aid me! But when women, who tend to convulsion fits, are given wrong guidance by the midwives or their children get fixed and lie athwart the womb in any of the many ways described previously and the midwife knows not how to manage it, it will be bad for the woman in labor. Indeed, convulsion fits have the power to push the child into an even worse posture. The pains press it down, but the convulsion fits work against the pains in the contraction and unusual contortion of all the limbs. I have at various times been fetched to a woman who had been lying for up to three or four days in a terrible delirium, and yet Our Dear Lord intervened so they all were delivered of their children with assistance and aid, as well as could be expected. Most of the women survived without a recurrence. Some of them, however, who were unable to recover, died a few days after the birth. This is a dire situation if the child lies wrong, inasmuch as the woman has lost her reason. Indeed, if the child merely lies crooked, it is hard to deal with unreasonable people; as long as convulsion fits last, they rob women of their reason and women resist help to their own detriment. So in such circumstances a child lying crooked and fixed can easily die, and the mother as well, especially if there is no midwife in attendance who can remedy the situation. If they are both dead, the blocked pains, together with the slippery putrefaction, vent. And so it is quite possible then for the slipperiness to allow the child to glide into the birth passage, and thus with the remaining moisture and the residual throes, that is, the pent-up pains, one thing drives another so a dead child can come forth from a dead mother, though it is unusual. I am of the opinion that a child born after the death of its mother must lie only a bit crooked and so can easily slip out, as I have just described. For if it lay really crooked or athwart the womb or if it were quite firmly fixed, it defies reason that it could be born after the woman's death. Thus it seems these women might easily have been aided by the midwife when this happened. Take this as a warning and as a point of instruction. You will see that with God's help you can save many women in this manner, unless it be that God does not wish it. Be that so, then all human ministrations and knowledge are for naught and are at an end. This has never happened to me, may God be eternally thanked! May God preserve me from this in future and help you, me, and all people in body and in soul, as may lead to our salvation!

CHAPTER 7

Of breaking the waters and how it is the responsible thing to do in dangerous births

CHRISTINA: I heard from you above that in some births when the skin around the waters is too tough, you can and must help by breaking the waters if there is to be a happy delivery and not a stillbirth. I would like to hear more about this.

JUSTINA: I can well recall various deliveries that gave cause for concern, deliveries that could be aided by breaking the waters to result in a happy birth. However, in certain cases where there was a failure to break the waters, it led to nothing but unhappy deliveries and stillborn children. Because there are some big differences in breaking the waters, you have to proceed with caution. Prematurely breaking the waters (I am calling it premature, but I mean when it is done unnecessarily and early on, since as a rule heavy labor ensues soon thereafter) would give cause for a bad conscience and would bring with it guilt here on earth and for all eternity. Likewise it is irresponsible to break the waters when a child is to be born at the appointed hour but lies wrong. For as long as the waters are unbroken, the child has room to turn to the right posture as it pleases. As I have observed and experienced, with the onset of the last pain before the waters break, it sometimes turns to the right posture. As soon as the waters break, be it naturally or by being broken, the pains squeeze the child as they find it into the birth passage and it has to stay that way thereafter, even if it is as bad as described previously. This disposition lies at the heart of most unhappy deliveries, and it cannot be altered thereafter except by turning the child, as I have already reported in sufficient detail. This causes not only painful and hard labor, but also the death of mother and child if no one can be gotten who knows how to turn the child. Thus, when breaking the waters, you have to think about what you are doing. If you do not have a good understanding of it, then let it be. For should you fail to break the waters at the proper time, how would you justify it? If the waters break naturally and a difficult delivery ensues, as often happens, you are not at fault. This pertains only to breaking the waters early on with children lying in the wrong posture. It says nothing about how irresponsible it would be to do it prematurely, which is, however, impossible. The waters cannot be broken until the inward mouth is open and the waters are there, for that is when the birth is at hand. It cannot be practiced without sufficient knowledge of the pregnant woman and without a little hook or some other appropriate instrument. Thus it is unnecessary to warn you against it and you need not take it on either. As irresponsible as the aforementioned breaking of the waters is, it is highly necessary in certain cases. I will show you how I gained knowledge of it with the personal testimonies of several women I have inserted below; I once had to request and acquire these for certain reasons and on account of malicious accusations.

CHRISTINA: How shall I grasp this? How is it that you too have been accused of breaking the waters prematurely and rashly?

JUSTINA: It happened because only a few people know it is possible, though not easy, to do it without an appropriate instrument. Before the waters press through the inward mouth into the top of the neck it is difficult and all but impossible to break them with your fingers. The waters cannot be broken merely by being lifted up and poked at, especially when the skin is tough; they recede rather than break. I became aware of that during the deliveries where Necessity demanded that the waters be broken if the child was to be delivered alive, as the testimonies that I got and included here—Frau Thym's and those of other women—will show you. And it cannot happen without the knowledge and consent of the woman in labor.

With certain complications I have encountered, however, Necessity has often required breaking the waters, but it cannot and should not be considered a premature delivery. I have been accused of this, but this accusation cannot stand up before Eternity; rather, it was a proper and speedy delivery at the appointed hour of birth when the pains were lasting and steady. The breaking of the waters mentioned here is done to prevent complications as, for example, when the navel string comes forth before the head when the birth commences, which puts the child's life in danger. When the throes come, the child suffers greatly because the navel string is crushed between the birth passage and the child's head. I have observed that if the birth occurs quickly and easily, such children are born live but very weak. If, however, the birth occurs slowly and with difficulty, then such children will be stillborn. This led me to introduce my hand halfway into the inward mouth, as soon as it was possible—in that the top of the neck will certainly tolerate [the advance—*trans.*] of the rest of the hand—and to break the waters so I could get right to the navel string as well as to the child's head. And I did my best to get the navel string back into the womb behind the child's head. I do the same thing even if the navel string has come forward but a little and I see it right away, having been present from the onset of labor. It can, however, never really go all that well on account of the thinness, length, and quickness of the navel string and the slipperiness of the waters, which makes it come forth fast. Nevertheless, the earlier you do it, the better; it is better to do it when the navel string has not long been out in front of the head. Whenever I am able to get it back inside quickly, I tell you, the children are delivered living and healthy, though the birth is somewhat difficult. On the other hand, if the navel string can be gotten back in only slowly and with difficulty, the children come into the world weak and sometimes stillborn, even though the birth itself was not all that difficult. When I realized this, I broke the waters as soon as possible and employed this means, namely, I smeared a soft, thin cloth with oil and with my left hand stuck it into the womb with a little spatula between the child's head and the side where the navel string protruded so the navel string could be held back. Worried about the cloth being left behind (which in fact never happened), I drew a thick thread through it, leaving one end hanging outside the womb so I could pull it out. I never needed it though.

Thus breaking the waters, when undertaken early on but at the proper time, also prevents a little hand from pushing out over the child's head. If the midwife becomes aware of it before the entire arm pokes out, she as a rule proceeds just as with the navel string. For if she breaks the waters, she can get to the hand unimpeded and it can easily be pushed back inside. As soon as the hand is out of the way, the head will enter the birth passage, which keeps the hand from coming out in front again. Since the waters drain off the child and Nature favors delivery by the head, I have often tried this in such cases, and happy births ensued. I hope you will become thoroughly acquainted with this so-called early, but nevertheless necessary, breaking of the waters by means of careful and considered touching. I advise you not to do it thoughtlessly or with undue haste or to make a habit of it so you will not let yourself be deceived. As much as it can be of use when done thoughtfully, it can cause injury if done thoughtlessly. It cannot be justified unless there be dire necessity and danger ahead, even if it causes no more harm than lingering labor. It would be desirable for no woman to have to suffer her waters to be broken, but rather for the waters and the child always to come forth at the same time, because this is the easiest and best birth. In that case no one would ever be subjected to malicious judgments, as I was.

CHRISTINA: I would like to know when it is possible to do it as you have described.

JUSTINA: Up to now I have spoken only of breaking the waters early on, for which I was condemned in an irresponsible and unchristian manner. Now I will tell you what is possible when the birth is underway. If at the proper time for birth the waters press downward, the inward mouth opens, and the throes persist, then it is possible to do it. When the mouth opens up and the waters press into the top of the neck, then it is always possible, but unnecessary in cases of children that lie right for birth.

CHRISTINA: I need to recall the navel string again because you said that when there is a lot of room when the children are not curled up too tight, the navel string can easily come forth in front of the child. Can the navel string come forth and endanger children in all the postures they lie in? Should you not always plug up the woman's birth passage with a towel so the navel string does not come forth first?

JUSTINA: You have peculiar ideas. You could certainly plug up the birth passage with a cloth so the navel string would not come forth from the womb, but it would not help the child if it lay wrong. The posture itself is the death of the child. No midwife would allow a child to be born like that if she knew how to prevent it. If she knows not how and lets it come in the wrong posture, the stopping up of the birth passage with a cloth to keep the navel string in the belly will not help. You can always plug up the birth passage, however the child lies, if this will help save the life of the child. The little scrap mentioned above is not for plugging up the womb or the birth passage, regardless of the position the child is lying in. No. Rather, if the child's belly is folded over, it always has a hidden hollow in it, as long as the child is alive, unless the belly should lie athwart the womb, as described above—in this case the child's body leaves no room. If, however, the child lies curled up and there is room, then the navel string must be brought through this space back up close to the child's body so the navel string is not crushed between the child and the birth passage, for the crushing of the navel string will rob the child of air, breath, and all vital forces. So the little cloth must be stuffed between the child and the birth passage where the navel string threatens to or could slip through. That not only prevents the navel string from slipping out ahead, but also its getting pinched, which is dangerous for the child. Simply sticking a cloth in the birth passage is an unnecessary act. Even if it does no harm, it also does not help unless it be put in the proper place. You must not use a towel, but rather a soft, thin piece of cloth. The navel string can come out ahead in all the postures, and it is generally a sign of a weak child. Nevertheless, it is not always dangerous if the navel string is poked back inside with dispatch.

CHRISTINA: I have to remind you again to tell me when breaking the waters is necessary. Explain it to me properly and in detail as you promised me you would.

JUSTINA: To my knowledge it is necessary when labor is underway with such bellies that are easily disposed to unnatural births. You should not think this means that by breaking the waters you can make natural births; rather, it means that by breaking the waters you can keep children in the right posture for birth. In capacious bellies, children as a rule easily turn to the right posture but then to the wrong one. It is also necessary to break the waters when the skin that encloses the waters and the child is too tough and cannot break without great force. Delivery is otherwise greatly hindered, and such a woman is put in the gravest danger because her strength is much sapped by unnecessary pains. With some bellies, the children—even if they lie in the right posture—quite often turn and come forth with a hand above their head (because the waters have remained too long intact, creating too large a space and expansion), and they get stuck on whatever they hit, such as the sharebone, with their hands pushing against it. As a result the child pushes itself back up, which happened in a case of mine before I knew how to prevent it.

It can also easily happen—as I also had to experience—that when there is this kind of room and there are strong pains, the navel string comes out over the head of the child, even if it does not do so right away when birth commences, as recounted above. This causes the child's death if it is not prevented, namely, by breaking the waters, as I explained before. For if the inward mouth is already completely opened up and if the child's head is in the right posture for birth, delivery is only delayed and blocked by a tough skin around the waters because the throes cannot force it to break. This injures the child as well as the mother. The child is worn out and born weak or even stillborn; likewise it can assume all sorts of wrong postures. I have often found it to be true when I was called in to assist other midwives that this and nothing but this was the cause of bad deliveries or stillbirths. For this reason I advise you, while weighing in advance all the circumstances I have laid out, to take this into account too and to prevent unnatural births when labor is fully underway. I do not insist on breaking the waters, but if the inward mouth is fully open, the child is right there in position, and I noticeably feel that the waters have not broken, then I consider it necessary, provided it can lead to no other danger than a delay in the delivery, not to mention that it can prevent the complications outlined earlier, since what makes it dangerous is the waters remaining intact too long.

CHRISTINA: Explain to me once again, so I can understand it even better, when it is harmful and irresponsible to break the waters.

JUSTINA: It is harmful and irresponsible to break the waters if you do it too early in the absence of the aforementioned complications since you have no reason to fear danger. It is not right, without there being dire necessity, to break the waters because it will lead to a dry birth and difficult labor is likely to ensue (even if not in all cases, it happens in many). It is best when the waters and the child come forth at the same time. So it is also irresponsible for the midwife to break the waters when she does not know what she is doing. If, for example, a child is not in the right posture for birth and the waters are broken, the child will remain as it is. For inasmuch as the waters leave enough room for children that lie right to turn when the pains come—as I have explained sufficiently—they can in lingering labor also make room for children that lie wrong to turn so they are in the right posture and a happy birth can ensue. If the waters have run off and they lie wrong, the dryness will cause them to remain lying in that posture and they will get stuck. This is the greatest source of danger for mother and child. Just as breaking the waters can keep the child that lies in the right posture in that posture for the birth, it can keep children that lie in the wrong posture as they are and cause an unnatural birth. If the midwife is imprudent when breaking the waters and does not know how to judge the situation, she will be just as imprudent in turning the child. Here I am reminded of a midwife in Liegnitz who undertook such an uninformed breaking of the waters in the case of a woman who knitted for a living, thinking that delivery was delayed only because the waters would not break and the skin was too tough. So she broke the waters before she even knew how the child lay. In the moment the waters came, the right hand and the right foot presented and came into the birth passage. The midwife was terrified and did not know what to do except put the hand and foot back in, as best she could. The delivery was thereby delayed for three days, during which time she used all her might to keep the child in the womb so it would turn to the right posture. This and the pains so sapped mother and child of their strength that the child died and the mother was just barely alive. As a result of extreme fatigue and weakness the pains ceased and the child remained in the womb because of the lack of pains. Neither the hand nor the foot could come into the birth passage anymore. So Nature took over: inasmuch as everything was stopped up down below, a prodigious vomiting ensued with nothing but a thick, black substance.[104] The midwife imagined she had done her part by holding things up and back. Since no more pains came, she was not needed for the time being; she just wanted to lie down for a while and recover because she was tired out from watching and attending. While she was sleeping, Fate would have

104. Siegemund works with the notion that when things are stopped up down below, something will emerge from the mouth, a common understanding of women's bodies in this period. Vomiting during labor is of course not unheard of. The blackness of the substance, if the substance was indeed truly black, could indicate that the lining of the stomach had ruptured as a result of retching.

it that I unexpectedly came by in a stagecoach on the way from Breslau to Liegnitz. Hereupon the good people bade me come to them, which thing I consented to do. I found the laboring woman at death's door and the child dead inside her. I was therefore afraid to examine her as I feared she could not bear my turning the child and I did not want her to die under my care. Her husband insisted in such a heartfelt and passionate manner that, because there was no other help at hand, I should try and see whether it might be possible to save her. So I went up to her, laid her down on a comfortable birthing bed, and found the child's right hand and foot bent in the upper part of the neck. I pulled out its little foot and searched for the other one, which was quite easy to find; it lay in the inward mouth fixed to one side toward the hip. I pulled it out like the other one, taking both feet together while pushing back the little hand, as best I could. I hung onto the little feet to pull out the child and in that very hour managed to get it out. Everyone was very happy, hoping the woman's life could be saved, but the vital forces she had lost prevented it, and she died several hours thereafter.

Here you see an example of when it is quite unreasonable and irresponsible to break the waters, as well as how stupid it is to push back the child's limbs. If the midwife had refrained from breaking the waters, both tiny feet would have come together because they were down below, and as a result of their being together, the child would have pushed down with its feet and drawn its little hand back upwards; the child would have had to sink right down, and its little hand would have gone back on its own. To be sure, it would have been an unnatural birth, for she had twice before had such breech or unnatural births, yet both the mother and the children came away from it with their lives. I am telling you this to warn you to use caution when breaking the waters, for breaking them like this is irresponsible. The child does not die as a result of your breaking the waters, but the child's wrong posture endangers both mother and child. If the child is in the right posture, it will not be injured if the waters are broken or if Nature causes them to break several days in advance of the birth, except that a dry birth or heavy labor will ensue. Yet breaking the waters itself does not endanger anyone's life. If you ponder this well, you can be guided by it.

CHRISTINA: If the waters break on their own and the fetus or the child is in a wrong posture, what can be done then?

JUSTINA: Then you need to look at my account of turning so you know how best to turn the child and how it will suffer itself to be turned. You are not responsible when the waters break naturally. God and Nature did it. I have spoken only of breaking the waters when Necessity demands it. Because I promised to show you some testimonials on breaking the waters, these now follow.

THERE NOW FOLLOW SOME TESTIMONIES CONCERNING BREAKING THE WATERS

We, the undersigned sworn officials of the municipal court of Liegnitz, Daniel Pitiscus and Christian Weyrauch, Jurymen; and Johann Friedrich Hildebrand, Bailiff of the Lower Court, do hereby publicly document and proclaim, wherever it shall be necessary, that on this date there appeared before us at the ordinary court the honorable and virtuous Frau Justina Siegemund, née Dittrich, the famous and experienced midwife of these parts. She testified by warrant in what manner she had been accused of some improprieties, in part on account of allegedly hastening birth, in part on account of breaking the waters, which she has always executed successfully. And because she needed certified testimony from various women whom she had served when they were with child as to how she had treated them in childbed, she requested that we take down their testimony in court. And so we did. And the following persons, who willingly appeared when informed of said Frau Justina Siegemund's request, were warned, as is customary, of the gravity of their testimony and not to testify as a favor to anyone or to injure anyone, but simply to tell the whole truth for truth's sake as to how Frau Justina treated them at that time. Thereafter all of them reported—by their Christian conscience with discreet words and proper modesty—expressly as follows:

1.

Frau Maria Thym, née Hermann, wife of Georg Thym, farmer, testifies to the following: before said time when Frau Justina first came to her and she made use of her services, she had had eight stillborn children, all of whom had presented wrong and unnaturally. When she heard that a woman had come here (this happened about twelve years ago) who could offer good and useful advice in such cases and such childbeds, she got the idea of calling on Frau Justina for counsel. Frau Justina was able to comfort her about her upcoming confinement until labor commenced, for she heard from her several weeks beforehand that the child was lying in the right posture. However, with protracted labor pains she discovered for herself from feeling the fetus reverse its position that a change was occurring in the birth. So she fetched Frau Justina and said, "Alas, Frau Justina, I feel a little something in the birth passage. Will it be with me like the other times?" Thereupon Frau Justina examined her and found that the child lay in the wrong posture. Its little hand was in the birth passage and in fact without the waters having been broken and with the inward mouth completely dilated. Upon discovering this, Frau

Justina desired her to lie down so she could see whether it were possible to get the child into a position to be delivered properly. So she suffered it to be done. She then managed to reach beyond the hand to the child with the waters still unbroken and brought the child's little head before the birth passage. As soon as there were labor pains, however, Frau Justina had to let go of the infant because of the tension of the waters. So it returned to its previous wrong posture. Frau Justina then pulled it before the birth passage a second time, but it fell back once again when the labor pains came. Frau Justina tried it a third time and guided the child toward a proper delivery. The waters broke around ten in the evening, whether as a result of the touch or naturally, she did not know. Thereupon she labored the entire night with much moaning and groaning, but the birth would not ensue. So everyone demanded to hear Dr. Kerger's advice on the subject since Frau Justina knew of no other means but to push back the head and search for the child's little feet, but did not want to do this without informing Dr. Kerger first.[105] She feared that the child lay rolled up too tight and thus could not be delivered; meanwhile, as the pains continued, the inward mouth closed up completely. Hereupon Dr. Kerger was fetched first thing in the morning.[106] He advised against the midwife's intervening in the birth in said manner and maintained that the pains were still too weak for giving birth. He said he would strengthen them with physics so that she could be delivered, dead or alive. So despite the fact that they all begged him at least to await the effects of the first powder, he left on account of his other patients, claiming that he could not offer much help for the moment under such circumstances and that the powder would do its work and promote delivery, be she dead or alive, and that they would know to inform him further. When she got the powder and used it, she felt such pains coming over her that she nearly lost her reason and was completely black and blue. She only heard this from others since she herself could not recollect anything and had nearly lost her reason. As she lay in childbed, it appeared that she would suffocate if she were not raised up and brought to a sitting position on the birthing chair. The effect lasted over two hours so that she was sapped of her strength and could not possibly take the other powders without endangering herself in a way that gave cause for great concern. Upon this turn of affairs, she, at her own peril, beseeched Frau Justina to save her and to deliver her of her child in

105. Dr. Martin Kerger (d. 1691), town physician of Liegnitz, who became Siegemund's archenemy. See Siegemund's description in chapter 8 of her trials and tribulations with a physician whom she calls Sempronius.

106. The German reads "with the opening of the gate," presumably referring to the opening in the morning of the town gate, that is, first thing in the morning.

whatever manner possible. Thereupon Frau Justina got her to lay down her head and upper body so that she could introduce her hand to seek the child and push the fetus back by its little head. It was very painful but did not harm or injure her. Frau Justina immediately found the child's little feet beneath its chin and was of the opinion that the child's lying rolled up in a ball was the reason why it could not be delivered. Unfortunately a stillbirth followed, making it her ninth.

It thereafter came to pass that God blessed her once again, and indeed she asked for Frau Justina once again. She had her examine her about once a week, whenever she came to town, since she was constantly fearful about the child's posture. She heard from her six to eight weeks before the birth that the child still lay in the right posture. But four weeks before her delivery she fell ill with a consumptive fever that affected the fetus as well, so the child remained quite small. When she went into labor, Frau Justina thought that because the child was small it would not have the strength to turn and assume a wrong posture, as had the other children, but she was mistaken. Indeed, things turned out differently in that this child, like the previous ones, turned when the violent throes came and had to be delivered by its feet. But because it was small, it survived, was baptized and lived to be thirty-six weeks old.

Furthermore, when she thereafter summoned Frau Justina to her for the third time, the child, like the others, had early on presented for a natural delivery and stayed like that until the onset of true labor pains. Meanwhile, however, Frau Justina had seen the two other times how the children did not turn until the onset of violent pains. She blamed the child's reversing its posture for all of the unhappy deliveries and told them frankly that she knew of no other way to keep the child in the right posture than to break the waters, to which they all consented.[107] So upon the onset of mild pains she did it. Whereupon the mild pains ceased. True labor pains did not commence until the third day but then rapidly so none of the neighbors could arrive in time. She was delivered of a healthy daughter, who is still living and is now eight years old.

During her next pregnancy, she without hesitation once again sought out Frau Justina to assist her in the confinement awaiting her, and Frau Justina consented to do so. However, shortly before the delivery Frau Justina got a carriage and horses that were to take her to Brieg.[108] Nevertheless, she examined her at her request in order to see whether the child was in the

107. The testimony does not identify the "they" referred to here. Presumably these are the parturient mother, family members, and women attending the birth.

108. Town on the Oder River in Lower Silesia, now Brzeg in Poland.

right posture this time. Thereupon Frau Justina comforted her with the news that the child was lying quite right, that it would not be more than three days until it came, and that nothing was lacking except for someone to break the waters, but she did not want to do this in case she unexpectedly had some other kind of worrisome complications during the delivery and since she herself could not attend her this time on account of the trip she had planned. It came to pass once again that during protracted labor the child turned and after hard and heavy labor was stillborn the third day after Frau Justina's departure for Brieg.

On the occasion of her last childbed Frau Justina humbly asked Her Highness, the duchess of sainted memory,[109] graciously to give her leave and was thus with her during labor. At the onset of mild pains she once again broke the waters. Thereupon the birth was likewise delayed for three days. A living, but weak child was delivered and this past New Year's Eve died of a severe case of smallpox. Moreover, she testified on Frau Justina's behalf that the latter had practiced her profession honorably with her, had treated her carefully, and had not done her the least injury, etc.

Tantum[110]

2.

Frau Regina Titz, wife of the honorable Gustavus Rothusang, citizen and cabinetmaker, testifies as follows:

She had conceived seven children with her husband and had used Frau Justina as midwife for six of them. In the first two deliveries the children were in the wrong posture from the beginning until they came forth. Never mind that Frau Justina had been present during her confinement as much as possible and had tried to put these children in the right posture. The second time (the first time the child was already dead when Frau Justina arrived) she felt with her own hand Frau Justina's hand up under her ribs on the right side and noted how she tried to bring the child's little head down beneath its body. But it was impossible. The child was delivered feet first and was stillborn. Fourteen days before the birth her third child had in fact turned so its head was in the right posture for delivery and stayed like that until the day before the delivery. Then, during the night, her pains commenced, but she did not think these were actually labor pains. The following morning she had Frau Justina fetched, but it was already too late because the fetus had

109. Luise of Anhalt-Dessau had died April 25, 1680 (see above, note 28).

110. Latin: that is all.

turned over the course of the night with the onset of the pains. No intervention was possible, and so it had to be delivered feet first and was stillborn.

The fourth was also delivered feet first and stillborn. Frau Justina was not present for this one.

The fifth, which Frau Justina once again attended, also did not turn for birth, and like the others, was delivered feet first and was stillborn.

She used Frau Justina once again for the sixth child. Because she had not come there[111] until the day before, as soon as the pains commenced, she immediately had her summoned to her. And so she found the child lying right for birth. And because she had realized while attending the third birth that a child that lay right could turn, she offered her the choice of breaking her waters as she had with Frau Thym who thereby gave birth to two children. She said she knew of no other way of saving the child and was much afraid of the child turning as had the others. So she suffered it to be done, and three days later she had a happy delivery and a living child who, as long as God were willing, remained lusty and healthy and on next April 19 would turn three years old.

The seventh child did not turn to the right posture for birth and thus, like all the others, was delivered feet first and stillborn. For this seventh birth she at first used a different midwife, named Frau Corneliuß, because Frau Justina had advised her to try a different one to see if it perhaps would go better. Then things got dangerous, and she had to fetch Frau Justina after all. And she states and confesses in good conscience that in all these deliveries not a single part of her body was ever injured by Frau Justina, and she has no complaints to press against her; on the contrary, she has to thank her, next to God, to this very hour for the life of her only child, etc.
Tantum

3.

Frau Barbara Jentsch, née Geißler, the wife of Georg Jentsch, citizen and cooper, reports that she can in good conscience say that in her most recent two confinements Frau Justina, with God's help, saved the life of her children. After lingering labor and an unnatural birth with her second child, she had forfeited her health as a result of a fallen womb. So she had been led to entrust herself to Frau Justina for the third delivery. Upon suffering strong pains and going into labor and recalling her first childbed, she demanded to

111. The testimony does not identify the place and indeed, unlike most of the other testimonies, mentions no place names whatsoever.

know whether the infant was in the right posture for delivery. Frau Justina answered that she must not be worried; everything was as it should be. Thereafter Frau Justina was called away to Frau D. Volckmann with the consent of the witness for a mere quarter of an hour. When Frau Justina came back and brought her to the birthing bed, she discovered when she touched her that the child had turned during this quarter of an hour and was in a wrong posture with its hands and feet before the birth passage. To be sure Frau Justina then attempted to get the child in the right posture. This was impossible because the pains were too great. It was thus delivered stillborn, feet first, as had been the first child. Frau Justina had treated her with such skill in her confinement that she recovered her lost health.

When God blessed her once again, now for the fourth time, she asked again for Frau Justina. Frau Justina examined her various times prior to her confinement to find out whether her children could turn in and out of position while she was still on her feet as it were because in her opinion with such a belly it would be imperative to break the waters. And the child was in the right posture until the time of delivery; when the pains started, she had her reservations about relying on the child because in her previous childbeds in the beginning the infant had been in the right posture, but when the pains became violent, it turned. Thus she and her husband resolved, in God's name, to allow Frau Justina to break the waters because she said that doing so was the best way to save the child. And she did it, and a happy delivery followed inside of two hours. The child lives still, hale and hearty and sound of body, and is now in its ninth year.

When God blessed her yet again, she asked for Frau Justina once more. Eight days before her delivery she went to Frau Justina herself to find out whether the child was in the right posture; it was discovered not yet to have turned. Six or seven days later, because she did not feel well, she went to Frau Justina once again to be comforted. She learned then that the child now lay in the right posture. But Frau Justina sent her away with the warning that as soon as she felt herself going into labor she should fetch her promptly because children who turned shortly before birth seldom remained in the right posture when the pains came. In such a case she must not delay breaking the waters to keep the child in the right posture. Thereafter she suffered pains the entire night, but thought that it was not yet serious, until three o'clock when she sent for Frau Justina to come to her. When Frau Justina arrived, she found her pacing the room, even with negligible and few pains; the child was lying in the wrong posture with its hands presenting and the waters still unbroken. Frau Justina was very frightened, but nevertheless took comfort in the thought that because the pains were not yet violent, she might yet be able to get the fetus in the right posture. Since the child had turned while she was lying down, Frau Justina had her laid in her lying-in bed, on a flat bed, to see whether the child might also turn to the right posture while she was lying down. Within an hour, with assiduous prayer and through divine aid, this did indeed happen. The child now lay just right. Because she was still having only weak pains while she remained lying down, Frau Justina could not possibly break the waters. So she had her get up, since when she was standing the pains got stronger. But when Frau Justina introduced her hand intending to break her waters while she was standing and the pains came, the child drew itself up and turned so it lay wrong as a result of the force of the throes and Frau Justina's attempt to break the waters. When Frau Justina realized this, she refrained from breaking the waters. She put her back to bed, hoping that the child might turn once again. This happened shortly. Thereupon she attempted to break the waters while she was lying down. When the throes came and she made the attempt, the fetus reacted to it by pulling back and turning upside down. So she could not undertake to break the waters while the pains were occurring. Then she got the idea of breaking the waters without the pains. She attempted this with a little wire hook with which she hooked the skin and tore it in two. This was successful, and on the same day toward evening, with very easy labor, a happy delivery ensued. This child is still alive and now in its sixth year. So she had to confess that, for her part, she had to thank Frau Justina, after God, for her own health, which she maintains to this day.

Note that the next two testimonies by high-ranking persons of quality are inserted here because they concern breaking the waters.

4.

I, the undersigned, do confess herewith to the following, wheresoever it shall be necessary. Because Frau Justina had been charged with improprieties in managing births and because she had delivered four children for me, she appealed to me and desired of me a testimony of how things went with these deliveries. Since I do not want to fail her and for the sake of promoting truth, I testify by God and my good conscience that I never heard such accusations either in Brieg or in the countryside, to which fact I, along with my relatives, can testify. Rather, everyone wants to have her—if only they could get her. With me she never did anything that was clumsy or brutal, neither by hastening the delivery (which I would in fact have liked to have happen before my lord traveled to *N.*,[112] had it been possible to do it in a good fashion) nor

112. *Nomen* (Latin: name): here a stand-in for a place name.

when she broke my waters, which she did during the third delivery, praise God. That child is still alive and is healthy and strong, a little boy. She did this with my knowledge for good reason. An honorable and wise woman, she always treated me accordingly. After God, I owe her enormous gratitude and have reason to praise the aid she gave me. I have signed this testimony with my own hand for greater security and have guaranteed it with my customary seal. Done in Brieg, on October 24, 1681.

L. S.
N. N.[113]

5.

I, the undersigned, do confess that when my daughter, Frau *N.*,[114] was with child, I requested Frau Justina to attend her in childbed. Here in our principality her aid is most sought after by pregnant women, and one and all seek her out so she is much in demand. So I sent for her very early on. I fetched her eleven weeks before the confinement to get her advice so we might not lose the child because my daughter was not doing well and we feared a premature birth. Frau Justina determined that the child was not lying in the right posture. She did not want to worry me, but she told my children that she feared it would be a difficult birth because the child would probably present with its feet rather than its head. And it proved to be so. Eleven weeks later she was fetched again and found the woman in labor in the throes of violent pains and in a perilous state. She revealed this to me and said she was worried that the child was not lying right and that it could hardly be expected to live since under such conditions scarcely one in a hundred could be saved, especially first-born children. The child lay sideways in the belly as it had previously when they had fetched her. Meanwhile the pains became stronger and stronger without letting up. The child would not let itself be positioned right for birth in one way or the other; what with all the pains, the situation was a dangerous one. Because I, along with the others present, perceived that the danger was great and feared to lose both mother and child, we fetched (*Tit.*[115]) Herr Dr. *N.* and (*Tit.*) Herr *N.*, our pastor, asking the one to stand by us with loyal advice and the other to comfort us and to see to it that if the child came out far enough, it would be baptized.

113. *Nomen,* here a stand-in for the signature.

114. *Nomen,* here a stand-in for a proper name.

115. *Titulus* (Latin: title). Both the title and the family name, represented here by *N*[*omen*]., were left out for greater discretion.

And because the situation became still worse, I, along with Herr *N.*, spoke to Frau Justina. After God, we had faith in her. We knew that she would stand by my daughter with every possible aid that she could justify before God and her conscience. Thereupon she offered to do everything that was proper for an honorable midwife to the best of her ability and to omit nothing that might help both mother and child, hoping that the frequent alterations might yet yield a proper birth. But if this was not to be, she wanted it to present with its backside (for the child was not as much in danger of losing its life in such a posture). She wanted to break the waters (she did not like doing that, but Necessity knows no law) so the child would not be able to slip aside and would have to stay in position. She did not know what else to advise. All of us gladly consented to this. A painful, but happy delivery ensued. Frau Justina showed herself to be careful, industrious, and laudable. So all those present lavishly praised her knowledge and art. After passing through slowly and with great difficulty, the child at first appeared to be quite dead. Its head had gotten stuck at the neck for a quarter of an hour. Nevertheless, with God's mercy and Frau Justina's loyal assistance it burst forth. To the great joy and astonishment of all present, mother and child were saved without any harm, defect, or injury. Praise and thanks be to the Almighty.

Because she asked me to testify to this, I did not want to fail to do it for her; rather, I wished to give this testimony in accordance with her request, and to the best of my knowledge and conscience. Brieg, May 15, *Anno*[116] 1681.

(*L. S.*)
N. N.

6.

We, Magistrate and Juryman of the municipal court at the royal chartered city Ohlau,[117] document and confess herewith, wherever it shall be necessary, that the honorable and virtuous and experienced midwife Frau Justina Siegemund, who is well-known in these parts, made known at length how she always honestly and uprightly assisted women in labor. She especially needed and requested testimony as regards breaking the waters, which she had often done—and to date happily in every case. Among others, she had

116. Latin: in the year.

117. Town in Lower Silesia, now Olawa in Poland, situated between the Oder and the Ohle River.

lent the honorable and virtuous Frau Barbara Stieff, our now-widowed paper maker,[118] a helping hand when she was with child in her marriage. For weighty reasons she had had to break her waters. We were pleased to call the same before us and interrogated her before the court as to how Frau Justina had treated her in her confinement.

Thusly summoned, Frau Barbara Stieff appeared willingly at our and said Frau Justina's request. We explained Frau Justina's petition to her, earnestly exhorting her, by God and her good conscience, to tell the whole truth and nothing but the truth in the interest of truth, with the intention neither of harming anyone nor of doing someone a good turn. She was to testify how Frau Justina treated her at that time. Thereupon, swearing by her good conscience, she recounted the following in discreet and seemly words. It would be two years come autumn that Our Dear Lord had blessed her with a child for the sixth time in her marriage. She had previously had five stillborn children and had placed her special trust, after God, in the oft-mentioned Frau Justina and her art. Frau Justina came then at her entreaty several days before her confinement. When she discovered the nature of her condition and learned of the five previous stillbirths, she thought it advisable on account of the disposition of her belly to break her waters. She suffered this to be done in God's name on account of the aforementioned weighty reasons, and so three days before her labor pains commenced, she had her break the waters, which Frau Justina did with such dexterity that she not only felt well and was in good health thereafter but three days later gave birth to a young, healthy daughter without a single *obstacula*[119] or prolonged labor pains. This little daughter remains healthy, so long as it please God. With this well-executed maneuver, she certainly proved herself an upright midwife. She thus had to say to her credit that she was careful and cautious with her during her confinement, indeed, that she had treated her in such a way that, after God, she could never thank her enough.

When said Frau Barbara Stieff resolutely stuck to her testimony, the oft-mentioned Frau Justina Siegemund requested a certified testimonial of what was just reported. Thus we herewith grant it to her in documentation under the ordinary seal of our municipal court. Thus it happened at Ohlau, October 25, 1682.

118. Stieff is identified as the paper maker, not her husband; she must have taken over the business upon his death.

119. Latin: obstacle. Note the use of Latin in the paraphrased testimony; most women did not have the opportunity to learn Latin in the seventeenth-century German-speaking territories and therefore would have been unlikely to use such Latin terms.

There follow here once again the testimonies from the Municipal Court of Liegnitz.

7.

Frau Maria Lorentz, née Hütter, wife of Heinrich Lorentz, miller, along with Frau Elisabeth Hütter, née Scholtz, wife of Baltzer Hütter, tawer,[120] testify to the following. Three years ago, on May 3, when she, Frau Lorentz, was taken sick in the middle of the night and already had a midwife in attendance and it was found that the child was presenting with one little hand and the navel string, Frau Justina was called in to help her. When she arrived (at around four o'clock in the afternoon), she examined her at her entreaty and determined that the little child had definitively entered the birth passage with its left hand and left side—the midwife had been pushing its hand back the entire afternoon. Thereupon Frau Justina called upon everything she knew. The woman in labor was already very weak and lay at death's door, and the child was, as Frau Justina reported, already dead and giving off an evil smell. In this state of things the child's little head was impossible to get, especially because the waters were already entirely gone and it was to be a completely dry birth. Frau Justina claimed to know of nothing else to do but search for the child's little feet and thereby save the mother's life. So Frau Justina introduced her hand into the birth passage and pushed in her arm up to the elbow and then said the following: "This is extremely difficult, but as I have never left a child in its mother's belly, God will stand by me here too." She asked for a straight slender rod and a cord, three ells long. So Frau Hütter brought her a candle dipping dowel, along with the cord, and at her request made a notch at one end of the dowel and wedged the cord into it. Thereupon Frau Justina took the dowel with the cord wedged into it and threaded it between her arm and the dead child up to her own hand, which was still in the womb. She did not know what Frau Justina did with it in the womb. Thereafter the child, though dead, was finally brought into the world. Afterwards, when the child lay in the basin, Frau Justina showed everyone how it had lain in the womb. So she could say in good conscience that, with God's help, Frau Justina had saved her life. Her health had suffered no injury from this; after a year had passed, she gave birth to a healthy child that was now nearly a year and three quarters old.
Tantum

120. Person who converts skin into white leather by mineral tanning, as by soaking in alum and salt.

8.

Frau Barbara Vogt, née Gartke, wife of Georg Vogt, the leaseholder of the game preserve of the estate, testifies to the following. It happened nine and a half years ago when they were renting the Wittich farmstead before the Glogau gate on the Töpfferberg[121] that at three o'clock in the morning on a Saturday she went into labor. Thereupon she had Eva Kern, who was at that time well-known on the Töpfferberg, fetched to her as midwife. Because Frau Kern deemed herself to be too lacking in understanding to make a judgment, she requested that an additional midwife be fetched. So she had Frau Corneliuß fetched to her that very same day along toward evening. When the latter arrived, she pushed her so hard with strong medicaments and exhortations to bear down as hard as she could that the birthing chair broke beneath her. And because she had still not been helped, on Sunday around noon Frau Wittich sent the sworn midwife, old Frau Maria Mäuer,[122] out to her in the carriage. She returned along toward evening without having accomplished anything. The first two midwives stayed with her until Frau Justina came to her on Monday evening, having been requested to do so. She had not been able to make herself available earlier on account of other women in travail. Frau Justina was of the opinion then that the child was in the right posture for delivery. After spending two hours without getting anywhere, she said that if she were content to have her do it, she wished to touch her with her entire hand to determine what was wrong, which thing she suffered to be done. Upon touching her, Frau Justina determined that the child lay with its neck before the birth passage and was already dead. Thereupon Frau Justina laid her down upon a flat bed with her head down. She pushed the child back by its shoulders and raised its head so it could enter the birth passage. Frau Justina persisted in this for two more hours, hoping that the birth would ensue by these means. But it was nevertheless to no avail since the child's neck was broken completely in two. Knowing of no

121. Gate in the city wall on the north side of Liegnitz toward Glogau (Polish: Glogow), a city on the Oder River in Lower Silesia, now a part of Poland. Töpfferberg, literally "potter's hill," became a part of Liegnitz proper in the nineteenth century.

122. The text identifies Maria Mäuer literally as "the sworn old Frau Maria Mäuer." It is not clear in what sense Mäuer is sworn. I have chosen to translate the word as "sworn midwife," as is most likely the case. It is, however, possible that Frau Mäuer was a so-called sworn woman rather than a sworn midwife. In 1687 Völter characterized sworn women as women who stand in for the official midwives when the latter cannot fulfill all of their obligations. He pointed out, moreover, that in some regions they were entitled to collect fees (Christoph Völter, *Neueröffnete Hebammen-Schul/ Oder Nutzliche Unterweisung Christlicher Heb-Ammen und Webe-Müttern.* 2d ed. [Stuttgart, 1687], 78). Note that Frau Mäuer serves Susanna Jacob as well.

other means, Frau Justina asked for and was given a cord several ells long with which to put a loop around the child's head. She could not get the loop over the entire head, but only to just over the child's eyes, where it cut into the bones, but the loop would only stay put when it had been drawn up around the neck. So Frau Justina protected her bladder so carefully from the knot in the loop by putting her finger in front of it that she forfeited and lost the nail of her middle finger. But all the effort Frau Justina made to aid her was to no avail because the cord caused the child's head to be pulled back into the same position Frau Justina had originally found it in. Thereupon she desired Frau Justina just to leave her be; she wished to die. So all three of them left her alone and lay down for a while and fell asleep from exhaustion. So she could now do as she pleased, toss and turn as she wished, thinking that she could thereby help herself. In the process the blood in her belly was so roiled that she tasted it in her mouth. Finally, when she could no longer hold it back, she threw it up. This happened several times. Because this exhaustion had brought her close to death—so she felt—she asked Frau Justina once again for help, begging her to try one more time to save her life. So Frau Justina finally resolved to hook the child. But there was no appropriate crotchet at hand. Eva Kern had brought along her weeder, which was a sort of hook. Frau Justina raised the child's little head once again, as reported above, and pushed the hook into the child's head up in front at the fontanel. So taking the hook and the loop together and with the aid of a number of accompanying labor pains she was able to extract the child and deliver it successfully. As soon as the child was extracted, her head felt quite good again and all her pain had vanished, and she was thereafter brought to bed to rest. And since then she had had no need to complain of difficulties with her womb or of any other problems with her belly; since then no part of it had caused her further pain, so she had to thank Frau Justina, after God, for saving her life this time.
Tantum

9.

Frau Susanna Jacob, née Ritter, etc., testifies voluntarily with good will and conscience, and is willing in any case to back it up with a corporal oath: in the year 1673, on November 15, around noon, she called upon Frau Justina to assist her. She had gone into true labor on Saturday, so she thought, and for this reason had had Frau Mäuer fetched to her as midwife. Frau Mäuer had recognized that she was in full labor and had prepared to deliver her. She labored with violent throes from Saturday until early on Tuesday when

all her pains vanished, and around noon a violent hiccupping ensued. When the pains stopped, Frau Mäuer left her and went home, saying she knew of nothing more she could do for her. The child would probably perish. They should call her. When it was time, it would come. When her hiccupping continued, she had Frau Justina fetched, who found her alone, except for her female attendant, and in dire straits as a result of the hiccupping and increasing nausea. She thought that she could not continue any longer in this manner without endangering her life. Frau Justina examined her to determine whether it was possible to help her and told her that she could be helped, but that she should fetch Frau Mäuer so as to prevent a quarrel, since she had been with her for several days. As soon as Frau Mäuer arrived, Frau Justina spoke to her in the following matter: "Examine the woman and see whether things are as they were when you left her or whether something has changed in the meantime and she perhaps could now be aided." Whereupon Frau Mäuer replied that things were as they had been for a long while, that they would have to await the proper time, and that she knew of no way of helping her. A quarrel arose because Frau Justina said it was necessary to help her and that she could be helped and Frau Mäuer maintained the contrary. Finally she had to mediate between the two of them and begged Frau Justina, for the love of God, to save her, which she did, and before Frau Mäuer's very eyes, she helped her so well that she was delivered in two hours of a young daughter who was still living and was now ten years old. Neither her belly nor the child were injured in any way during the delivery. She thereby concluded her testimony.
Tantum

10.

Frau Maria Müller, née Roth, testifies with good will and in good conscience and maintains that she can certainly strengthen her testimony with a corporal oath: during her first two confinements she was in danger of losing her life because the afterbirth remained behind, especially in the last and second childbed. She labored from August 1 to August 6 in the year 1677, that is, for five days, until she was saved with great effort by Dr. Kerger and Frau Justina. In the year 1680, on March 8, her life being in danger, she tearfully begged Frau Justina with all her heart to deliver her third child, as she had promised, insofar as it was her only hope. But because Frau Justina had unexpectedly received a command from the widowed Duchess of Brieg[123] around eight to fourteen days before her confinement, she reluctantly had

123. Luise von Anhalt-Dessau (see above, note 28).

to let Frau Justina go. Nevertheless, she asked her upon her departure to give her good advice about what to do if she once again were imperiled as a result of the afterbirth being left behind in the belly. Which thing Frau Justina did. And she had to confess in good conscience that this advice served her well in her last two confinements, for which she will thank Frau Justina her whole life long, etc.

Tantum

After the witnesses had concluded their testimonies and had confirmed them with a judicial handshake they were, as is customary, dismissed *imposito silentio.*[124] However, Frau Justina was, in accordance with her petition, given the present attestation as documentation under the customary seal of the bailiwick of the lower court. *Actum*[125] Liegnitz, March 13, 1682.

L. S.

CHAPTER 8

Of Remedies

CHRISTINA: One more thing, dear sister. Tell me about the remedies you have found beneficial and necessary to use on those frequent occasions when things are rushed and it is not possible to get a doctor. Instruct me so I too can make use of them in these situations.

JUSTINA: You ask a lot of me and consider not what you ask. Remedies are also medicaments. They belong to medicine and not to our trade. In a case of dire necessity a midwife might recommend or try this or that, but outcomes show that good will is frequently the best thing; because of the unforeseeable complications and circumstances not everything serves everyone. The venerable doctors themselves often have trouble and reason enough to reflect; all the less do I wish then to have anything to do with remedies or offer instruction about them. In our introductory conversation I heard from you that you have already served some hundred women. While serving them you presumably heard about various remedies and how they sometimes work well and sometimes do not. The inconsistency of those who recommend them testifies to their ignorance. One suggests this remedy, the other that one; what one recommends, the other repudiates; and what this one has rejected, the other seizes upon. Indeed, I have learned that people reject what they themselves previously recommended. And even if some remedies might be beneficial, they are often not doled out right, but instead care-

124. Latin: charged with silence.

125. Latin: done (past participle of *ago*).

lessly, sometimes too much, then too little, given and used at the wrong time so they do not always have the desired effect, and thus those who recommend them must expect a lot of suspicion and lamentation, as well as a sharp reproach from the venerable doctors. Besides this, you already know how the innocent Titia was persecuted by Sempronius[126] on account of the remedies she employed. Regardless of the fact that she had learned about them earlier from Sempronius himself, she was accused of using coral liquor, spirits of shepherd's purse, and ruby water, that is, of using these three things in an emergency.[127] People have heard far and wide how said Sempronius tried to harm good Titia with the pretext of these remedies, which are otherwise common, customary, and reliable. The suit, which was dragged out over several years, made the remedies she had used in good faith all too costly for her. Considering then that Titia had to endure so much on account of remedies (which she in fact learned about from a highly learned old doctor and which were in and of themselves beneficial and commonly used), you would be much less secure with my instruction. It is better then to learn from her misfortune. I advise you against devoting yourself to or relying on remedies; it is much safer to seek out the venerable doctors in a timely fashion and let them worry about it. If in an emergency no one can be reached, then make cautious use of what God and the occasion provide you and what you know can do no harm.

CHRISTINA: It never would have occurred to me that there would be so much to say about remedies. But since you have reservations about granting my request, tell me how good Titia reached a resolution with Sempronius, for I have heard people discussing it in noble houses but have yet to learn the outcome.

JUSTINA: Since a Christian should uphold the honor and good name of his neighbor and not keep silent about what concerns the innocence of his neighbor, I will reveal to you briefly how Titia solidly proved her case against Sempronius so he finally had to be contented. I have all her files in my possession and read them often and thus am thoroughly familiar with them.

126. Sempronius stands for the very same Dr. Kerger mentioned in testimonies 1 and 10 in chapter 7. See the Volume Editor's Introduction.

127. The common weed shepherd's purse was believed, among other things, to stay bleeding and was variously tied to the loins, the neck, the backs of the knees, the armpits, and soles of the feet to regulate menstruation and stop the uterus from bleeding. Coral liquor was used to treat epilepsy, melancholy, bloody dysentery, and hemorrhaging of the uterus. It was thought to cleanse the blood and to fortify the stomach. Ruby water was used to revive the life forces and to strengthen the nerves as well as to ease the pains of colic or maladies of the uterus. It was normally administered in the form of compresses laid upon the pulses.

Titia was a sworn midwife in an important city, and said Sempronius was her good friend. After a few years he let himself be misled into no longer being well disposed toward her (I will not cite the reason). For this reason his previous good will turned into all manner of torment, which led Titia to seek retirement from her office. The town let her go at the displeasure of the authorities, but with a lovely testimonial to her good conduct. Practically all the inhabitants, particularly many noblewomen in the countryside, lamented her departure and complained about her hardship. She returned to a safe private life under exalted princely protection. Sempronius was glad to have banished Titia from his sight, but it was still necessary to drive her out of the hearts of those well-disposed toward her. For this reason he gave the honorable town council of one and the same city a paper in which he cited against her the use of the above-mentioned remedies and other dangerous things. Not, to be sure, that he wished to accuse a person in absentia. Rather, because many women in the city still asked for Titia, he only meant to enumerate these things so the honorable town council would no longer allow her to be given the occasion to come there, indeed, more importantly, so it would completely cut off her access. Precisely this would likely have come to pass behind the scenes since Sempronius had spread through said paper, as well as by word of mouth, all kinds of alarming and suspicious warnings among women of quality. But by Divine Providence her praiseworthy services and sympathy moved some loyal people to inform Titia about all this. Titia soon after demanded to see the paper that had been submitted against her. Among other things she discovered that the alleged informer had suggested to the honorable town council that when Titia came to town she should be taken into custody. So Titia came and exchanged a number of papers with him and got instruction in the matter from three universities, and when these came in in her favor, she demanded a legal judgment in the case, which was being drawn out, inasmuch as the honorable town council certainly recognized that Sempronius had gone way too far and after the passing of so many years had alleged things that were untimely and unfounded. Despite the dubious nature of all this, Sempronius resolved to line up excessive testimony against Titia and rounded up sixteen witnesses—noblewomen, bourgeois, simple country folk, midwives, even young girls who could know nothing about childbirth—thinking Titia would be drowned by the flood of so many witnesses, the authorities deafened, and the suit delayed, or that among the sixteen witnesses there would certainly be something said during the testimony that he could make use of and that at the very least could render her suspect. Titia paid no heed to any of this, but insisted instead that each witness be examined individually under oath and with a great deal of effort pulled it off—which outcome Sempronius could probably never have imagined; thereafter she demanded to see the testimonies, which were kept from her. Why? Because (as the successful outcome proved) all sixteen witnesses had testified in Titia's favor and against Sempronius and Sempronius thus could hardly prevail. In the meantime Titia managed to acquire a post in her trade abroad, yet she did not rest until she managed to see to it that the testimonies of the witnesses were sent to her. When she examined them, it became crystal clear that Sempronius himself had dictated the testimonies to the noblewomen. However, he had persuaded the bourgeois women to testify on his behalf by badgering them, or he interpreted their private remarks otherwise than intended and to his advantage. He forced some of the midwives, who were under his supervision and thus feared him, to say bad things according to his instructions. Indeed, he had not hesitated even to cite witnesses who were to testify that Titia's service had been harmful to them, when Titia, in the case cited, had not even been there, but instead had been on a trip. In sum, all sixteen testimonies made clear in a single voice that Titia had shown the witnesses much benefit and loyalty. They had nothing but good and kind things to say of her, etc. And it is not common that sixteen witnesses examined under oath testify in favor of the person they have been cited to testify against. God wished to prove Titia's innocence to the consolation of all midwives.

CHRISTINA: God help us! What am I hearing? Shall I believe it or not?

JUSTINA: Dear child, if you do not believe my story, then believe the lovely testimonies Titia gave me for safekeeping since she often has to travel to distant lands. At the end of this conversation in the first part I will show you the three testimonials, certified in print, from the most honorable electoral and exalted princely universities on the subject of the above-mentioned coral liquor, shepherd's purse, and ruby water. So you will see and learn that Sempronius's denunciation was nothing but harassment and that it is not even possible or humane to practice what he otherwise accused Titia of doing. Should it be necessary or if someone should question her innocence, I can show you the large work in writing, the written testimonies of the witnesses, certified by the court, or, indeed, I can easily have it published.

CHRISTINA: Dearest sister! Simply allow me to continue and ask once more what else Titia did and how it went with Sempronius when Titia received these cherished testimonies and official statements of her innocence! For I want to know how the matter ended so I can learn from it how to preserve my good name just in case, inasmuch as we midwives are subject to many annoyances and none of us can know what will happen to us over time.

JUSTINA: When Titia (as I said before) received the testimonies of said witnesses, she sought guidance from eminent scholars. Some advised her to carry out the suit and prosecute Sempronius. Others, and indeed the most important among them, offered this piece of advice: she should write to the authorities before whom the action was pending that they should give Sempronius a legal deadline by which time he should put in writing what he thought he had proven against Titia with his many witnesses. When Sempronius could not be brought to answer, Titia was further advised to prove her innocence with the testimonies. So she had an excerpt prepared from the testimonies and in it at last requested justice against Sempronius. She sent this document to the appropriate place, and a universal silence followed. Thus it is not proper for me to say anything more about what may have happened with Sempronius. It was easy to advise Titia, however, who was conciliatory and did not want to seek to disadvantage her neighbor; indeed her lord commanded her to attend to her present employ since the testimonies were more in her favor than otherwise and since she was everywhere trusted and esteemed. Her trade did not permit her to carry on any further protracted law suits in other lands, for which reason she could not be allowed, as she had requested in a petition to her exalted lord, to travel there herself and to prosecute her case in person. So it was incumbent on Titia to be obedient and pleased, inasmuch as even with a pending case, she mainly aimed to preserve her innocence and to commend her cause to God. She never showed any desire for revenge and sought nothing more than to preserve a good conscience before God and a good name in her profession before the world. She preferred backing off from the right she had in hand in order not to bring Sempronius still more trouble. She was fully of the opinion that he would recognize his own hastiness before God and would beg God's forgiveness for all the annoyance and expense he had brought upon her. Indeed, her attitude toward him was so Christian that she testified to various people that she forgave him from the bottom of her heart, recognized she had become more famous and renowned as a result of this trial, and had seen with her very own eyes that all this torment served her best interest and led her to patience and reconciliation. And she had all the more reason to thank God, who revealed her innocence. Since that time God had richly blessed her in her profession and shown her mercy and granted her greater insight in many and difficult births. That is why she had even less cause to hate him and hold a grudge. Rather, she was compelled to ask God to forgive him and bless him further in his profession and to preserve in her this Christian and forgiving disposition to the end of her days, which end she is now becoming ever more prepared to meet. And this is the true report of the case of Titia and Sempronius where remedies constituted the primary cause. So one more time: do not resort to and rely too much on remedies. Practice your profession faithfully and assiduously, and if you pay attention to everything that comes to pass, you will see and hear a sufficient amount about remedies, but in the end you will find that no clever midwife is so legally privileged that she would not be subject to lawsuits as others are. Rather, God will intervene and protect her so persecution may come to an end, and this holy vocation will maintain its unimpeded course to His honor and the furthering of the human race. This kind of thing came to pass here, praised be He.

CHRISTINA: I thank you a thousand times over for your report. It encourages me and all oppressed midwives and shows how they should defend themselves when persecuted like that. May God reward you too for your conscientious lessons on difficult births and the prevention thereof when possible and on adroit turning of children who lie wrong! May He bless you to your dying day and let your name live on after your death. May I grasp your lessons well and practice them fruitfully to honor Him and to benefit my neighbor! Preserve me and all midwives from envious, devious, and malicious enemies, mend the ways of those who already exist, and be merciful to us all! I still wish to read the three testimonials from the faculties of medicine you promised me

JUSTINA: I will gladly grant your wish so you will believe all the more in Titia's innocence and see the unpleasantness she experienced on account of remedies.

RECOMMENDATION OF THE PRAISEWORTHY FACULTY OF MEDICINE AT THE UNIVERSITY OF THE ELECTORATE OF BRANDENBURG AT FRANKFURT ON THE ODER ON ACCOUNT OF CERTAIN REMEDIES IN THE PROFESSION OF MIDWIFERY ETC.

P. P.[128]

Inasmuch as we saw from the paper left with us that you desire our judgment on some questions that you sent us, we, Dean, Senior Professor, and other Professors of the Medical Faculty at the University of the Electorate of Brandenburg at Frankfurt on the Oder, do answer them in the following manner:

We are asked first, whether coral liquor can be used to combat the lassitude of newborn children who exhibit no other weakness. Therefore we do

128. *Praemissis praemittendis* (Latin: sending in advance what is to be sent in advance) a locution that replaces the lengthy salutations and titles in rough copies of official documents in the seventeenth century.

answer that in this case a midwife would not be wrong to use them. The second question: whether a midwife can be justified in binding spirits of shepherd's purse to the pulses when a woman suddenly begins flooding in labor or when she is aborting, or if it is not available, in using ruby water instead of liquor of shepherd's purse; to this we do answer: no error can be imputed to the midwife for doing so. The third question: whether a midwife can at her discretion peel off the secundines[129] of a pregnant woman before it is time for the birth without harming mother and child; to this we do answer that such peeling off of the secundines cannot occur without injury resulting soon thereafter. The fourth question: whether a child that was removed prematurely, that is, a child whose birth was hastened, can live into its third year and whether if it thereafter dies of a certain illness, the acceleration of the birth must have been the cause of death since no other symptoms had been noted previously; to this we do likewise answer that this is not possible.

To corroborate this more emphatically we have confirmed it with the seal of our faculty. Frankfurt on the Oder, June 13, 1681.

(*L. S.*)

Dean, Senior Professor, and other Professors of the Faculty of Medicine at the University of the Electorate of Brandenburg at Frankfurt on the Oder

From this you see, dear sister, how a midwife can get into trouble using beneficial and proven remedies. The following two judgments from universities will enlighten you further about the case that Sempronius submitted with regard to opening certain veins, breaking the waters before the time of birth, and peeling off the afterbirth early on to hasten the birth and the harm that could thus be done to the health of mother and child. His case was fundamentally refuted.

REPORT TO THE UNIVERSITY OF LEIPZIG

P. P.

Our office obliges us to report to the same[130] that Sempronius, a surgeon in this town, alleged *per modum denunciatonis*[131] in court of the communal midwife

129. Secundines: the placenta and fetal membranes expelled from the uterus after childbirth.

130. That is, whichever officials at the University of Leipzig are being addressed by *P. P.* (see note 128).

131. Latin: by way of accusation.

Titia, who otherwise is much praised for her skill and much used by the nobility in the countryside, (1) that she hastened the delivery of pregnant women by a certain opening of the veins so that she could travel from one woman to the next as suited her and arrive everywhere in good time, (2) that she habitually and prematurely—and before the disposition of pregnant women otherwise called for it—broke the waters for the purpose of a regular acceleration of birth, (3) that by peeling off the secundines early on in the eighth month she further promoted the birth so that women too would be relieved of their burden.

Titia took this accusation for an *atrocissimam injuriam,*[132] and upon examining the matter thought to corroborate the accuser's *animum injuriandi*[133] from the fact that everything she was accused of was in and of itself impossible and absurd because (1) she knew nothing of such veins and the surgeon could also not specify them; at the very least they were not situated in a spot where they could be opened for such a purpose by a midwife without the most awful flooding or even causing the death of a pregnant woman; (2) it would also be quite impractical to break the waters prematurely because this never ever led to the acceleration of the birth, but instead merely *imminente & se notante partu*[134] habitually served to keep a fetus that was in the right posture for birth from turning again; (3) that was a notorious impossibility and absurdity because with *praegnantibus*[135] the inward mouth was shut tight and hardly ever dilated before the appointed hour of birth. Hence it would have to be opened with great force to peel off the secundines, which could not happen without the woman suffering the most excruciating pain and the fetus being endangered, even if a midwife were able to penetrate that far with her hand. Moreover, the child's head, if it were in the right position, would lie so close to the inward mouth and the sharebone that if one wished to reach the secundines, one would have to push it away and turn the natural state of the fetus topsy-turvy. Furthermore, the liver cake[136] or the thick flesh of the secundines was attached up above in the fundament of the matrix and everything was enclosed by a delicate skin; it was impossible to peel off the secundines without injury to these unreachable things and without breaking the waters.

132. Latin: most awful injustice or injury.

133. Latin: intention to do [her] wrong; a threatening or announcing in advance the spirit of doing wrong or causing injury.

134. Latin: in the case of imminent and self-evident birth.

135. Latin: pregnant women.

136. Placenta.

Because this juridical matter cannot be decided without a discussion as to whether it is possible to do what Sempronius accuses Titia of, it is incumbent upon us *pro communi Bono*[137] to inquire whether the *partus*[138] is precipitated through such injurious manipulations and whether harm can thus be done to the health of mother and child.

Thus we request our highly honored and most gracious lords to ponder this at length according to the *fundamentis Medicinae & Anatomiae*[139] and furthermore to inform us *specifè*[140] for the sake of truth, since Sempronius alleges of yet another midwife that she practiced these arts in another town a few years ago, but Titia gainsays this as *absurda & impossibilia*. For which special favor we are in your debt and prepared to compensate you, we remain etc.

ANSWER OF THE MEDICAL FACULTY AT THE UNIVERSITY OF LEIPZIG

P. P.

Inasmuch as the same[141] demanded our *judicium Medicum*[142] in a letter sent to us, dated September 15, on account of a denunciation submitted by Sempronius, a surgeon, *contra Titiam Obstetricem*[143]:

Whether the *partus* is precipitated through such injurious manipulations, like those attributed by Sempronius to Titia, to wit, opening certain veins, breaking the waters before the hour of birth, and prematurely peeling off the *Secundinae*[144] and whether harm can thus be done to the health of mother and child?

Thus we have found after lengthy deliberation according to the *fundamentis Medicinae & Anatomiae* that this allegation of Sempronius *contra* Titia is quite erroneous and false because delivery cannot readily be promoted by mere bloodletting and even less at will, inasmuch as there exist not a few examples in which frequent bloodletting did not do pregnant women the least

137. Latin: for the common good.

138. Latin: birth.

139. Latin: fundamentals of anatomy and medicine.

140. Latin: in detail.

141. That is, the addressee, designated here by *P. P.* (see note 128).

142. Latin: medical judgment.

143. Latin: against the midwife Titia.

144. Latin: secundines (afterbirth).

harm. And Sempronius needs to specify the veins that Titia is alleged to have opened and how often and in *qva copia*[145] she bled her. In addition, he needs to name what symptoms occurred in the *Gravidis*[146] thereafter. It is also to be noted with the breaking of the waters that pregnant women's waters often break a good long time before the delivery on their own and that they suffer no harm from it, although sometimes a hard birth ensues. Besides, on account of the inward mouth, which is shut tight, it is not possible for midwives to get in far enough to break the waters before the inward mouth dilates and before the waters present and thus before imminent birth. It is still less possible to peel off the secundines, especially at the spot where they are firmly fixed by the *placentam uteri*[147] to the matrix.

Which thing we wish herewith to report for your information under our seal. Leipzig, September 23, 1682

(*L. S.*)

Dean, Senior Professor, and other Doctors and Assessors of the Medical Faculty here

TO THE PRAISEWORTHY MEDICAL FACULTY OF THE UNIVERSITY OF JENA

P. P.

We herewith kindly give the same to interrogate in what manner Sempronius, a surgeon, some time ago entered a denunciation here with the office of the court marshal of the Electorate of Saxony, which office has been graciously entrusted to me, a denunciation against a midwife, whom we shall call Titia, who otherwise is praised far and wide for her experience and skill. This denunciation was on account of three outrages she is alleged to commit habitually, to wit,

1. She allegedly conducts an impermissible and premature intervention, entering through the vagina and the inward mouth (*per vaginam & osculum uteri internum*[148]) into the cavity or hollow of the fundament of the matrix, and there she is said to open up certain internal veins of the womb or at least to try to find some veins that could be opened by

145. Latin: what quantities.
146. Latin: pregnant woman.
147. Latin: placenta of the uterus.
148. Latin: by the vagina and the internal os of the uterus (that is, cervix).

penetrating the matrix through the vagina up to the inward mouth in order in this manner to hasten the delivery of the pregnant woman so, since she is summoned to many exalted personages, she can cover everything and will perhaps not have to neglect one person or another.

2. She allegedly habitually breaks the waters prematurely *ante ruptionem Amnii ordinarie solitam*[149] and thus makes her work go faster in order to hasten the birth and get away as soon as she wants.
3. She habitually—and this is the most irresponsible and dangerous thing of all—controls the beginning and advancement of the delivery as she pleases by prematurely peeling off the secundines, even already in the eighth month or occasionally later, to the extent that she can, if and when it should please her, go from one pregnant woman to the next and thus deliver women without being detained a long while.

However, inasmuch as the midwife Titia took these accusations that Sempronius had made against her—accusations that can be found entered at length in the records—to be nothing other than *atrocissimas injurias,* she deemed this denunciation to stem purely from the would-be accuser's slanderous disposition. She thus maintains that everything that had been launched against her, as mentioned above, is, on the one hand, completely impossible, and on the other hand, also contrary to all *principia anatomica*[150] and the practical experience of the art of midwifery. For she asserts

Qvoad[151] (1) It certainly cannot be denied that the anterior veins, through which the monthly cleansing of women takes place as a rule, open up sometimes in certain cases when the midwife touches a woman or even without the examination.[152] However, it should not be immediately imputed to the midwife that she has opened the veins that promote birth and thus has prematurely promoted the birth, *idqve eo minus,*[153] even if her hand is stained with blood when she removes it on account of running into these open veins.[154]

149. Latin: before the normal and usual rupture of the amniotic sac.

150. Latin: anatomical principles.

151. Latin: as to the point that.

152. Menstruation was commonly understood in the seventeenth century to occur as a result of veins in the uterus opening up to rid the body of excess blood.

153. Latin: and that even less so.

154. The controversy cited here probably centers on the show, that is, a mixture of blood and mucous, that typically occurs shortly before or after labor commences. Patients can mistake this show for bleeding (Margaret F. Myles, *Textbook for Midwives with Modern Concepts of Obstetric and Neonatal Care.* 8th ed. [Edinburgh: Churchill Livingstone, 1975], 243). Indeed, as Siegemund describes, if a hand is inserted at that point in childbirth into the vagina, it may come out bloody.

Seeing as no such veins, whose opening can promote the delivery of a pregnant woman, are to be found, the surgeon and the would-be accuser would never ever be able to specify any. For if he immediately cited the arteries and blood vessels in the placenta or the so-called liver cake, he would soon be taught otherwise by the fact that flooding never ensued and the fact that the rapid death of the fetus followed even less so. Nor could he mean the veins in the navel string because these are all encased by a membrane as well and likewise cannot be opened, or rather torn to pieces, without killing the fetus immediately.

Qvoad (2) She stated, however, that she only employed the breaking of the waters if the fetus had assumed the *situm partui proximum,*[155] or, as she expressed it, had presented well, so that it did not assume a bad or dangerous *positur*[156] by turning in a way that gave cause for worry, which often comes to pass. This, however, certainly does not mean promoting a premature birth, but rather doing it at the appropriate hour, all in good conscience.

Qvoad (3) And finally as far as peeling off the secundines is concerned, Titia says—inasmuch as this appears to her completely absurd—that she cannot imagine that a sensible *Anatomicus,*[157] as this surgeon and would-be accuser claims to be, should affirm and allege this sort of absurdity, seeing as how the *osculum cervicis uteri,*[158] or the inward mouth, is so tightly closed with *Praegnantibus* that it hardly ever opens before the appointed hour of birth. Thus she could not even dream of introducing her hand and forcing open the inward mouth, as it were, thus causing the pregnant woman the most dreadful pain, indeed even putting the pregnant mother, along with the fetus, in inevitable danger of losing their lives. Thus she claims never to have undertaken to peel off the secundines. And if the would-be accuser wished to think and counter that she did it when the fetus had already penetrated the birth passage, then he would never be able to demonstrate credibly the possibility that she could get her hand in past the head of the child that had penetrated the birth passage, considering that once again flooding and the unavoidable breaking of the waters, indeed extreme life-threatening danger for the pregnant woman, as well as the fetus, would certainly occur in the process, whereas in contrast the latter had never yet occurred, praise God, through any highly improper neglect attributable to her. Instead the examples cited by the surgeon, that is, the would-be accuser, all testified in her

155. Latin: position for imminent birth.

156. Latin: position.

157. Latin: surgeon.

158. Latin: cervical os.

favor, in that there were two-and-a-half-year-old children among them, as well as several who were thirty weeks old, who had reached a reasonable age in good health and some of whom were still living.

Since I would like to be informed in detail about all this, I herewith submit my friendly petition to the entire praiseworthy faculty for them to consider at length all of these points and the circumstances cited: whether Titia's assertion is truly the case in anatomy and the art of midwifery or whether what Sempronius accuses her of actually takes place and whether it is possible to undertake manipulations like those he cited against Titia with the said effects. Above all (especially because this is very important to the entire country), I wish to inform myself with a detailed and well-founded report *in Medicinā & Anatomiā*. I am in your debt for your obliging accommodation of my request and will not fail to render payment of the fee *pro labore*[159] with my gratitude at a suitable moment. In the meantime then I remain, sheltered in Christ's mercy, ever,

The well-disposed friend of your praiseworthy faculty
Dresden, April 6, 1683

ANSWER OF THE PRAISEWORTHY MEDICAL FACULTY AT THE UNIVERSITY OF JENA

P. P.

From Your Excellency's gracious petition, we have observed at length in what way a surgeon, Sempronius, came forward *denunciando*[160] against Titia, a midwife, on account of various outrages, on which matter our report based in *Medicinā* and in particular *Anatomiā* is desired.

After assiduous collegial perusal and weighing of each and every alleged circumstance, we have found that the *denunciatio Sempronii*[161] rests on shaky grounds, whereas Titia's response indeed corresponds to Nature, the tenets of medicine, and experience. Thus we do not hesitate herewith to confirm truth and innocence in the interest thereof. For even though

1. It is not groundless to assert that a great deal of harm can be done to a woman in labor by means of premature and untimely pushing and ex-

159. Latin: for the work.

160. Latin: in an accusing manner.

161. Latin: Sempronius's accusation.

amining, in that the fetus is thereby inhibited from participating itself and turning naturally; the woman giving birth, however, is exhausted from laboring vainly and prematurely. But it does not follow from the circumstances cited that Titia did this, but instead that she waited for true labor and the dilatation of the cervical os, inasmuch as it is more than certain that bleeding starts and stops several times on its own, not only before the birth but especially afterwards, even though everything has gone well, and that the bleeding as a rule bursts out in great quantities and thus has to be stayed.

2. Since Nature itself makes way for the fetus when the birth commences by breaking the membrane in which the fetus is enclosed and releasing the waters, thereby heralding the birth, careful breaking of the waters thus serves such hastening of the birth when the fetus has entered the birth passage more than it serves to cause harm, and therefore nothing untoward should be imputed to Titia, for she did this at the right time, and since it is often necessary, will continue to do it.
3. As far as prematurely scraping off the afterbirth is concerned, this is, in and of itself, impossible, as has been touched upon sufficiently in the report in that these secundines—inasmuch as it has been alleged *per impossibile*[162] that they can be peeled off—are much too firmly tied to the fetus and the matrix, and it would be completely absurd to say that premature dilatation is possible at will and that you could as you please pick the fetus like a piece of fruit from the stem without any resistance.
4. That you can find certain veins in women that can be opened so they give birth is certainly false and beyond comprehension, inasmuch as it cannot be the veins of the matrix that customarily undertake their monthly cleansing when the woman is not pregnant, since in such condition these are more closed and hidden; still others pertain to the fetus. That is why it is proper to find Titia innocent in this case too, as it becomes clear from everything that has been reported that bleeding that starts on its own with some women, even without external intervention or force, produces sufficient indication of the commencement of dilatation.

Documented with our faculty seal in Jena, April 24, 1683.

Dean, Senior Professor and Professors of this same Medical Faculty

162. Latin: as is impossible.

CHAPTER 9

Of tipping women in difficult births and the presentation of a comfortable birthing chair or bed

CHRISTINA: Those are really lovely and incontrovertible testimonials; they not only clear up Titia's innocence with regard to Sempronius's accusations, but I, along with all wise midwives, get thorough instruction from them as to the extent to which we should use remedies and observe the practice of breaking the waters when called for, since such highly learned people judge it not to be improper but rather approve of it and acknowledge it as beneficial and necessary if it is done with right and proper caution in emergencies. So no reasonable person can write or say anything against it.

JUSTINA: That is precisely why I wanted to prove with the testimonies I included earlier in what manner I too have over the many years of my practice and experience often found breaking the waters, when called for, highly beneficial. You could fill a big book with them. But reasonable, peace-loving souls will recognize sufficiently that my teachings rest on a solid foundation, and I do not doubt that they will accept them with gratitude and employ them, as God wills it, for the good of their neighbor, to which undertaking the Almighty, to whom faultfinders are also subject, will grant His blessing and mercy. God will graciously stand by those who read said instruction thoroughly, consider it carefully, and practice it cautiously, as He has me, and He will accommodate them with praise and honor.

CHRISTINA: Dear sister, I will keep the lessons you have given me in my thoughts and my grateful heart! I ask you not to take it amiss that I abuse your kindness once more. Do tell me in conclusion what it means to tip women in these difficult births. I have often heard talk of it. When do you tip the woman? Before the waters have broken or when they break? Or when the child is already dead and the mother in danger? I would think if women could be helped by tipping, it would be easier for the midwife and the afflicted woman than turning the child as you described before.

JUSTINA: You can imagine that if a woman laboring in a difficult and unnatural birth could be helped with tipping, I would have shown it to you previously without your asking since I have often had to come to the rescue, especially of peasant women who had labored three and four days long in sore travail, when the women had been tipped more than too much. Because they saw it did not help the woman, they finally summoned me, and then, next to God, turning the child provided the best aid. You desire to know what tipping is and when it is necessary: when the birth is just beginning or when the waters have broken or in extreme danger? I can easily inform you as to what tipping is. The woman is thrown over or, as I call it, overturned. There are various views on this. Some tie the woman to a board and turn her on her head; some roll her to one side; some lay her on the table and toss her from the table onto a bed of straw so she is overturned while in the air, as it were. But all this is dangerous. Misfortune can thereby easily befall the laboring woman. She is more likely to die than to be thereby aided. This tipping does not take place except when the child is dead and the woman in danger of dying (according to all appearances and in the opinion of the women attendants). So you can imagine how tipping is supposed to help when the child is dead and the mother is in extreme peril. This danger does not ensue when the waters are still intact, for as long as the waters are present, mother and child are not in danger of losing their lives on account of the birth. Children do die in their mother's belly when there have been no pains, but this dying comes not from difficult births (in my opinion) but from a random internal sickness. Pregnant women also often die from various illnesses. I am speaking here of the fact that in difficult births or with children lying in the wrong posture there is no danger of the child dying as long as the waters are unbroken; when, however, the waters break, danger arises, but sooner with one birth than with another, depending on how the child lies. And as long as no danger can be perceived, no one thinks of tipping. Meanwhile the waters run off and the child becomes dry and gets squeezed in so it is difficult and, even with good sense and proper knowledge, hardly possible to use your hand to guide and turn it. So you can easily perceive what good tipping will do. If it is that difficult to turn the child with your hand, then the woman could be tipped a hundred times over if her strength held out and the child would nevertheless remain stuck in its press, however it lies. Thus tipping is done in blind ignorance and originates with foolish people who think because the woman is tipped over, the child will also tip over. They do not comprehend that the child is stuck fast. Stick a piece of meat in a sack, tie it up tight, turn over the sack as much as you like, untie it again and you will find the meat is in the same position as it was when it was tied up, even if the sack has been overturned a hundred times over. It is the same with tipping. The woman will be suffocated before she will be aided by such absurd assistance.

CHRISTINA: But what if she were tipped earlier on, namely, when her waters were still intact? Perhaps she could thereby be helped and the danger averted.

JUSTINA: Tipping cannot take place when the waters are intact. It would likely make for unnatural births, but it would not make unnatural births right. That is not possible because the child would push up right against the pelvic bones (or the locks as they are otherwise called). If the child is to be turned to the right posture for birth, its head has to push into the pelvis. Tipping cannot bring that about if it does not come to pass naturally and by Divine Providence. Thus in this you see God's grace and omnipotence, that is, if you consider rightly how children are born happily each day. This view of tipping seems to me like the superstition many people have about knots on a pregnant woman and who therefore untie hair ribbons, apron strings, garters, and whatever she has on her that is tied, mistakenly thinking she will not be delivered of her child until all these knots are untied. This is superstition, and it is necessary to disabuse foolish people of it. For we believe that God rules over birth and has everything in His hands. What should these miserable ribbons or knots delay? In this view they would have to stop God. That would be an unchristian view. But those ribbons and knots that are tied too tight, especially around her body or her legs, and that would cause the woman pain could certainly be untied on account of the discomfort, but not on account of the superstition that the woman as a result of untying ribbons or knots will be delivered sooner. Far be it from me and you and all true-believing Christians to think that! Thus I advise you to pay good heed to the lesson I give you. There you will learn what really goes wrong in deliveries, what is to be done, how you can help with God's mercy and blessing, and how your skilled hand should govern it physically. Now to close this first part I will also present to you a comfortable birthing bed in a copper engraving and explain it in detail with figures; this bed is necessary and profitable to use in difficult births, indeed, in all births at your discretion, especially for refractory women who want now to sit, now to lie. May God bless your intention and mine with His grace that it may accrue to His honor and the good of our neighbor.

Presentation of a comfortable birthing chair or bed:

1.1.1.1. The four supports of the bed
2. The sideboards
3. The headboard
4. The foot end
5. The back, which you can prop up so you can lean on it like a backrest and which can be propped up on an iron rod behind the headboard of the bed and inserted in the holes in the headboard made for that purpose, high and low—as you like—and which works well to aid

the woman. You can put it all the way back until it lies against the headboard of the bed.

6. The seat,[163] where the woman sits

7.7. The armrests and handgrips she can hold on to. Take a look at the special armrests indicated by 7, how they are to be fastened onto the sideboards.

8.8. The footrests on both sides of the bed

9. The board that is inserted into the footrest and that can be pushed back and forth to make it short or long so the woman can put her feet on it, depending on whether she is large or small; you can see on the special footrest indicated by 9 how the little board where the woman puts her foot is held fast by a peg and nails.

10. Foot warmer that is set on the floor of the bed where the feet can be placed when the laboring woman wants to sit as on a birthing chair. Then the footrests against which her feet are braced are lifted out and put aside if it is too crowded. This makes the bed both a proper birthing chair and a birthing bed, however you want it.

11. The handgrip, where the woman can hold on tight with her hands, is also especially indicated. This handgrip has to be inserted into the armrests in various holes when she is lying or sitting so the handgrips are always comfortable for her.

Now look at it from the front:

1.1. The supports; 5. The backrest of the birthing chair, which you can prop up or put down; 6.6 is the seat and an insert that can be slid back and forth beneath the seat; so when the woman wants to sit, you push the insert back and the cutout in the seat is exposed and open to aid the woman; if she, however, wishes to lie down, you push the insert back beneath the seat so the cutout in the seat is shut again. Thus the woman's back lies firmly braced and in a comfortable position for labor. 7.7 shows the armrests, along with the handgrips; 8.8 are the footrests where the ridges and holes are so the little board (9.9) can be slid back and forth; 10 shows the cutout in the floorboard. The midwife must sit on a comfortable stool in this hollow beneath and with the woman, as 11 indicates, so she is half in the bed and half out of it and so she can back up and move forward.

Look at it from behind as well:

1.1 are the supports; 3 is the headboard seen from behind; 5 is the backrest along with the iron rod of the birthing chair you can prop the backrest

163. The German word refers to a cutout seat like a toilet seat.

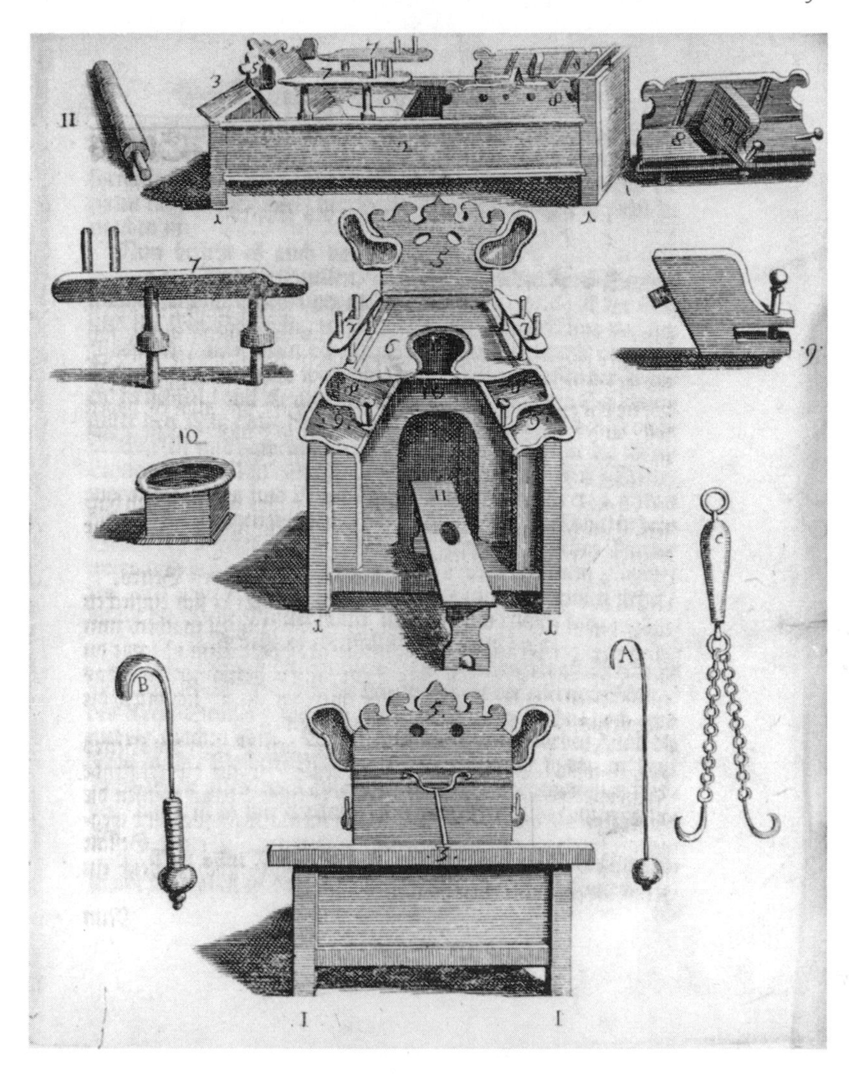

up with, high or low. When the birth is over and the woman is laid in her lying-in bed, everything that was set up in the birthing bed can be removed. Then you can lie in it like any other cot.

The open footrest is made with an insert, as can be seen above, which is indicated by 4 in the bed that stands to one side. It can be lifted out and slid back in.

The seat with the cutout in it is also made so it can be removed. However, it is affixed to the floor of the bed with pegs so it will not give way in the bed; it can be conveniently lifted out and put back in.

You can also remove the backrest.

I have commonly used the following three hooks:

"a" is the one I used in the beginning with common people when I had no better knowledge of instruments like this.

"b" is the one I found easier to use after I gained further experience.

"c" is a tidy invention for double hooking, which, however, I have never found necessary to do.

End of the first part.

PART TWO
CONTAINING A TEST TO SEE WHETHER CHRISTINA HAS PROPERLY UNDERSTOOD AND GRASPED JUSTINA'S LESSONS

First question

JUSTINA: Dear sister, now that in the first part I have at your request kindly and faithfully revealed to you my knowledge and experience of difficult births and how to prevent them when possible, as well as how to turn children who lie wrong, it will in turn surely not displease you if I now test you to see whether you have understood and grasped my lessons in all the particulars. Such repetition can strengthen you in your profession and make you secure in it. Therefore, tell me, why do women have a womb?

CHRISTINA: Because they are the female sex.

Second question

JUSTINA: Where is the womb? Can it rise up out of the throat as some believe or can the intestines fall out through it?

CHRISTINA: The womb lies in the lower belly between the rectum and the bladder and on top it has a dome without an opening of any kind. It lies beneath the intestines in a special hollow and is held in place so firmly by the ligaments that it must surely refrain from rising up out of the throat. Because the womb has no entrance above, the intestines cannot fall through it or out of the sheath.

Third question

JUSTINA: Where does the inward mouth lie, and how is it to be sought and found?

CHRISTINA: The inward mouth is to be sought and found below in the so-called sheath.

Fourth question

JUSTINA: How can you locate and identify the inward mouth?

CHRISTINA: It can be found when you introduce the first two fingers of your right hand as deep into the sheath as necessary toward the rectum. It can be recognized as feeling like a nipple. The rest of the skin is completely uniform.

Fifth question

JUSTINA: Why is it necessary for the midwife to be familiar with the inward mouth?

CHRISTINA: It is necessary because knowledge of it enables you to determine the appointed hour of the birth, which must be done, to wit, whether the child lies right or wrong, the latter being sometimes preventable early on, if not entirely then in large part. Moreover, you can often keep infants in the right posture; in some bellies they turn over at the hour of birth, which leads to unhappy births.

Sixth question

JUSTINA: Does the inward mouth lie deep inside or right inside in one belly as opposed to another?

CHRISTINA: The inward mouth lies deeper inside in one belly than in another. In some bellies it lies very close in and is easy to reach. In the case of a fallen womb it comes out of the belly and you can see it.

Seventh question

JUSTINA: Is it closed up tighter and firmer in one belly than in another?

CHRISTINA: In some bellies it consists of very firm flesh and is closed up tight; in others it is very delicate and thin in consistency.

Eighth question

JUSTINA: Is it all the same for the birth whether it is closed up tight or loose or to be reached close in or deep inside the belly?

CHRISTINA: It makes a big difference in the birth whether the inward mouth is closed up tight or loose, as well as whether the mouth lies deep in the belly. When it lies naturally close up inside in a healthy belly so it can be readily reached and is closed up either tight or loose, the birth will be much easier. These are generally the happiest and easiest births.

Ninth question

JUSTINA: Does the inward mouth that is closed up tight and lies deep inside the belly present a danger?

CHRISTINA: Even if it is not dangerous, it is generally a difficult birth if these two things coincide. But it is sometimes quite dangerous with unnatural presentations in that these infants cannot possibly stay alive during such a difficult birth, whereas, in contrast, in the aforementioned ones children who lie wrong are born alive.

Tenth question

JUSTINA: What is the inward mouth? Can it always be reached at the hour of birth with one woman as with another?

CHRISTINA: The inward mouth is a plug or lock of the womb where the child is conceived, carried, and contained until natural birth. It has to be sought, as I said before, with the first two fingers of your right hand. Your fingers can clearly feel the difference between the mouth and the rest of the skin. One finger alone cannot do this because you identify it by encircling it. It is always reachable in the hour of birth. There are some, though few, pregnant bellies, where you can reach it only with great difficulty. This condition is to be found among very stout women or those who have large and unhealthy bellies so the fruit can tilt forward into the hollow of the belly in front of the pelvis. In that case the mouth rises up high and the neck seems longer than usual.

Eleventh question

JUSTINA: Why is it necessary to touch women in labor?

CHRISTINA: It is necessary to touch them to find out whether they are having true labor pains or not, whether the child lies right for birth or not, and whether it is time to labor or not.

Twelfth question

JUSTINA: When true labor pains occur and the child lies right for birth, does it do any good to touch the woman?

CHRISTINA: It is necessary and useful to touch her all the same, not only to determine whether the child will come forth soon, but also to make certain the infant is kept in the right posture and does not go off farther to one side than the other or fix somewhere. Touching also prevents a fallen womb. Indeed, a midwife can instruct a laboring woman by touching her as to whether or not she should push hard. Likewise, touching indicates to a proper midwife all the dangers that can occur in a birth. If she is forbidden to touch the woman, then she works blind, and a woman who forbids it can only blame herself for any misfortune that follows.

Thirteenth question

JUSTINA: That is what I think too, but do tell me how a midwife can inform others?

CHRISTINA: She can teach others sufficiently through the touch and by providing a good report on it. You have taught me that without touching, no one can learn anything.

Fourteenth question

JUSTINA: How can you test to see whether a midwife has good and thorough knowledge?

CHRISTINA: By the touch, whether she can give an account of it and answer questions about it. The experienced midwife can from her extensive practice give plenty of information because through touching she becomes familiar with all the conditions that portend happy or unhappy births—what the child's posture will bring and things concerning labor. If she cannot do this in detail, namely, tell you how the human belly is constituted at the time of birth and what complications occurred, how she dealt with these complications or difficult births, how she herself intervened and enabled the delivery, and how she had to help, it will soon become clear how you are to judge her. The stuff of a midwife lies not in mere talking or words. So she might claim, "I have had many difficult cases. Thus I am well versed in the fundamentals" (for there are many difficult cases and births that Nature itself governs without the understanding of the midwife). Rather the midwife must be able to recount at length what has to be done in and with difficult births. In this manner she displays the depth of her knowledge.

Fifteenth question

JUSTINA: You have answered in keeping with my view, but when is it most necessary to touch a woman: when the throes come the strongest or when they have passed or before they come?

CHRISTINA: It is best and most necessary for the midwife to touch the woman in labor before the throes come. The midwife can thereby better found her knowledge as to how the throes are coming, whether the infant's head is going straight into the birth passage, whether it is in the right posture, or what posture it is in, and what kind of complications are arising, which she cannot know when the pains are present, inasmuch as when the pains come, the waters expand so the skin is taut, like a bladder blown up to capacity. When the waters and the skin are expanded, as with pains that are commencing or already in full force, the midwife can neither feel the child nor determine whether it is in the wrong or right posture. The expanded skin often deludes the midwife, who thinks it is the infant's head, inasmuch as it is quite like the head in hardness and roundness. But when the skin breaks and the waters run off, they make the unfortunate discovery that the child is presenting with its little hands or feet, when the midwife had earlier done nothing but provide comfort, assuring them that the child was lying in good shape in the right posture. Touching during labor pains, which is commonly done, leads to that sort of mistake. Midwives are themselves at fault for this error because they mostly eschew touching without pains, whereas they would thereby learn more about what would and could happen during birth than if they did it when the pains were just starting or had already set in. It is best for a midwife to touch the woman before the throes come and gently persevere until the pains have passed. She can thus know how the child lay before the pains and all the better help it aright. She can also know how strong the pains are and whether the woman needs to be helped a lot or a little. Likewise she can know how the infant positions itself after the pains, whether it can change its posture or whether it remains in the right one, which thing, if properly heeded, is, next to God's help, a great means of preserving the life and health of mother and child.

Sixteenth question

JUSTINA: I am happy to hear you have heeded and grasped my lessons well, but do tell me, does Nature set a certain time for birth?

CHRISTINA: God has determined that Nature should set a certain time for bearing a child.

Seventeenth question

JUSTINA: Is it possible to determine or calculate this time down to the day and hour?

CHRISTINA: It is impossible to determine or calculate this time with certainty down to the day and hour for the following reasons: (1) God can unsettle certainty in whatever form and as He wishes so no person may glory in his own knowledge; (2) a woman often has her cleansing courses one more time after conception and yet is pregnant; (3) some have none of these cleansing courses as, for example, when they get sick and yet are not pregnant; (4) some conceive while sick like that so they know not when they conceived; (5) various women also have their courses regularly right up to delivery, though this is rare, and they are nevertheless pregnant—some irregularly up to the midterm of their pregnancy, some even more irregularly and yet are pregnant as well. How can the day and hour be determined, given such a state of affairs? (6) It is also impossible to calculate according to the quickening of the child because one woman feels it sooner than the other, depending on how strong the fetus is.

Eighteenth question

JUSTINA: Is there nothing you can observe then that will tell you when the birth is close at hand?

CHRISTINA: There are certainly signs that tell you when the birth is close at hand, down to a few days, but they vary and seldom occur in such a way that you could name the day and hour. If they do occur, God nevertheless keeps the hour to Himself. For the entrance of each human being into the world, as well as the exit, is known to the Omniscient God and not to mortals. Additionally, labor pains cause several hours' delay during birth if they are slow, or if they come fast, one after another, the birth occurs more quickly. Moreover, it can well be that if a midwife knows how to be guided by all the particulars, she can come close to predicting it, speaking from reason and extensive practice, that is, if a complication does not arise. But the vain assertion that you can predict it with complete accuracy is, in my opinion, an impossibility.

Nineteenth question

JUSTINA: Your answer certainly corresponds to my view, but is there then no observable difference between the final month and the preceding ones?

CHRISTINA: With all women there are certainly differences in the inward mouth in the final months, but with one woman more than another. A midwife can perceive the difference therefore if she assiduously heeds the last month, especially if she has attended the woman before.

Twentieth question

JUSTINA: How do you perceive such differences and how does the inward mouth change in the last month, close to the birth?

CHRISTINA: As far as I know, the inward mouth changes sooner with one woman and later with another. (1) In the case of bellies disposed to give birth easily, it changes not long after the midpoint of pregnancy, becoming soft. Also it tends gradually to become softer and looser when, around the eighth week before the birth, it proceeds to open slightly and little by little becomes more yielding. As soon as the opening sets in and the child has turned, false labor pains generally commence and the belly alternately expands. I have heeded this and noted in accordance with the complaints of many women that this opening became little by little ever more disposed to widen. The inward mouth can be so thin and pliant with those women who have easy births that I have often marveled at it. (2) In the case of those who have hard births, it remains rigid until the hour of birth is at hand. So the violent throes have to force it. This difference can be noted close to the birth, if you but heed it. In the case of natural births that are underway, the inward mouth, when it is so thin and pliant, opens perforce from labor pain to labor pain. If, however, it remains rigid, it will take time, because the pains have to force the whole thing. It has to be as wide open as the child is large. The midwife can and must be guided by this if she desires to have a proper understanding of birthing.

Twenty-first question

JUSTINA: What is necessary to determine whether a woman is close to delivery or not that could be interpreted maliciously by envious people and give rise to pernicious gossip?

CHRISTINA: Let envious and jealous people interpret it as they will; in our profession it is necessary for us to know this. I do not say this because I wish to or can with certainty reckon when it will be—it is for God, not humankind, to know the appointed hour of birth. But God has given us knowledge of natural things that can and must indicate approximately when the birth will be when it is close at hand. So I can reasonably speak of it, in keeping with your lessons. I do not reject this information, inasmuch as I use it to determine whether it is time for the delivery or not when a woman complains. For if a midwife does not have this basic knowledge, she is a bad midwife, and it can result in grave mistakes and indeed the death of mother and child.

Twenty-second question

JUSTINA: I am not talking about the knowledge that is so highly necessary to have during the delivery, but of the complicated conditions that show up during the final months.

CHRISTINA: It is proper to consider these, for they offer great insight before, in, and during the delivery and of course indicate many subtle changes and complications. I am well versed in this with God's blessing and your instruction. Women often experience complications that make them think they are pregnant. They are sufficiently strong in body and every day hope for delivery, but they are not pregnant—like what happened to you. If such blind midwives come upon them as they did you, a woman could lose her life in the process or at the very least her health. Thus you can see that the knowledge, which can provide a basis [for making a judgment—*trans.*], must not be rejected. After the midpoint, a teeming belly is easy to distinguish from an unhealthy one, no matter how strong the latter looks and feels; the changes in the inward mouth reveal the basic condition of a pregnant woman after the midpoint from one month to the next. For at the very least, two, even three months before the birth you can feel the fetus by means of the touch in most bellies, whereas feeling the belly externally can easily be deceptive. How often does it happen that one and another woman, even if she has previously had several children, deceives herself and thinks she feels life when there is none? All the more so can a midwife who is poorly versed in the matter be mistaken. So I consider it necessary to pay heed to all the conditions and symptoms that occur with pregnant women, as well as you can. Thus it is also necessary, if you wish to be careful in your profession, to take precautions early on with pregnant women who have been subjected to many unnatural deliveries, as you showed me at length in the first part. For in the case of such bellies, mother and child can be saved when the birth is commencing—if not in all cases, nevertheless in some—if the midwife is properly informed. So it is highly necessary for the midwife in such cases to attend to the condition of the child in plenty of time—I would say two or three months before the birth—so when the appointed hour of birth arrives, she will neither act tardily nor hastily. If these two points are to be tended to conscientiously, then it is highly necessary to learn how to recognize the imminent changes in pregnant women. If we desire to become familiar with these changes during birth, it is highly necessary for us to be familiar with the changes that occurred beforehand; otherwise we will not manage it and will be easily deceived in the true hour of birth, as often happens with ignorant midwives. Therefore this question was highly necessary, and thus no reasonable person can in good conscience speak ill of it. If they do, may they be commended to God. There will still be pious hearts who will accept it with thanks and serve God and their neighbor with it.

Twenty-third question

JUSTINA: I hope so too and must ask you further, however, how and in what manner you can recognize and distinguish so-called false labor pains from true ones.

CHRISTINA: So-called false labor pains only go crosswise and tense up the belly on top. The inward mouth shuts up tight and recedes upward with the cramp so it becomes quite rigid. As soon as the pains subside, the inward mouth opens up as it was before the pains. When, however, a natural birth is beginning and the child lies right for birth, true labor pains penetrate the inward mouth and from pain to pain gradually force it to open up. By touching you can quickly see how the inward mouth is opening up when the pains come. True labor pains are easy to recognize and distinguish from false ones.

Twenty-fourth question

JUSTINA: What causes midwives to proceed as if labor were starting and yet women afterwards go on for several weeks before giving birth?

CHRISTINA: False labor pains and foolish midwives are at fault; they have no basis on which to distinguish false pains from true ones.

Twenty-fifth question

JUSTINA: Can mere touching by an inexperienced midwife hurry Nature so a woman might be forced to give birth when it is not yet time?

CHRISTINA: I consider this impossible, given your instruction and my own experience; otherwise it would all too often come to pass. I have often been summoned to cases where the midwives had gotten labor going when it was not yet time and had kept at it for two to three days, at which time I had to tell them to leave,[164] and the women went round for several more weeks thereafter before they actually gave birth. It would probably be possible to manage it with force, but no force can be applied in secret. However silly or clever the midwife may be, the pregnant woman has to be aware of it.

Twenty-sixth question

JUSTINA: But can Nature be slowed down or hastened by an inexperienced midwife?

CHRISTINA: It is also probably not possible for the plain ignorance of the midwife to hasten or slow down birth, if she did nothing or knew nothing to do. For how many children are born with no midwife on hand? How often are they summoned tardily so in the meantime the child comes before the midwife gets there—as has happened to you as well! But where there is the slightest shortcoming, the midwife can fail to do many things so the woman is held up unnecessarily for several days. And even if it nevertheless turns out happily, mother and child can also suffer great misfortune.

Twenty-seventh question

JUSTINA: But may a midwife not assist where there is nothing wanting in the birth?

CHRISTINA: Although, when there is nothing wanting, children can and do come without the presence of a midwife, no woman can forego a midwife without fear. She must fear all manner of complications and thus always needs a midwife.

Twenty-eighth question

JUSTINA: Why does one woman give birth more easily than another? Does this come about from natural causes? How do you recognize these causes?

CHRISTINA: Everyone knows one woman gives birth more easily than another. There are, however, also natural causes of many difficult births that can be prevented or remedied with knowledge and skill, if not completely then appreciably. (1) The inward mouth can be at fault when it is so hard and rigid that it cannot give way as necessary for birth; (2) a difficult birth ensues when the inward mouth lies too near the anus and is pulled up high

164. That is, tell the midwives to leave.

by the fetus, which happens with bellies that hang out far in front; (3) a difficult birth ensues even when the child's head lies right for birth but has gotten lodged to one side or the other; (4) the birth is still more difficult and dangerous when a child presents for birth unnaturally, which can occur in various ways. You with your superior knowledge can answer these questions better than I. These complications can be somewhat aided with knowledge and skill, even, God willing, possibly completely prevented—all this in accordance with your instruction, which I have found to be true.

Twenty-ninth question

JUSTINA: Why is the inward mouth more yielding in one woman than in another?

CHRISTINA: Insofar as I have learned from you and my own experience, it is because the skin is tougher or more delicate with one woman than with another, and the more delicate one gives way sooner than the tough one. For this reason the weakest women often have the easiest deliveries, whereas the strongest ones have the hardest ones. Age can play a large role in it, but here too it varies, as you explained at length in the lesson you gave me.

Thirtieth question

JUSTINA: When you have one of those ill-disposed bellies that is to give birth for the first time, the woman is no longer young, the child large, and the woman herself stout, can a happy birth result as long as the child lies right for birth, and how does it come out if the child lies wrong?

CHRISTINA: These deliveries are difficult and practically the most difficult among all deliveries for mother and child, even though the child lies right for birth. And you have to manage such deliveries gently and thoughtfully. Nevertheless, you cannot always save the children because they easily suffocate on account of the hard, slow birth and because of their size. Indeed, often the mother must also forfeit her life, as you lamented earlier, on account of the arguments among the midwives and attendant women, for such deliveries can be brought about in no other way than with a crotchet if the child is dead. They will sooner neglect the mothers before they allow the crotchet to be employed. When the child lies wrong, however, people seldom object to turning it; indeed, their own eyes tell them there is no other way. You can also get hold of the child better by the feet than by the head, and in such a state the woman is sooner saved.

Thirty-first question

JUSTINA: What if the child's head is too big?

CHRISTINA: This delivery must be handled carefully and very gently, and you have to pay good heed to the natural pains so you do not lead the woman to push too hard. You simply must not strengthen the pains; otherwise the large head will be pressed so it is wide rather than pointy. The pains generally appear to be weak and brief in such a state. They are, however, weak and brief solely because they cannot rapidly force the child until the child's head has become pointy.[165] The pains then ensue on their own, and a happy birth follows. If the woman is encouraged to push too early and too hard, it can result in the death of mother and child.

Thirty-second question

JUSTINA: You answer right and proper, but how can you determine when a child lies right for birth and yet has its head fixed to one side? And how can you help it in that case?

CHRISTINA: A child lying right for birth can easily be recognized by the hardness of the cranium and by its roundness, likewise by the open head at the fontanel, if the waters have not yet broken. And a head that has fixed, whatever side it be on, can be felt quite well and dislodged as soon as the waters break and the pains have passed. Before another pain comes, you must insert the two fingers usually used for this up to the child's head. You gently span its head until the pains come, at which point it is very easy to guide it away from whatever side it has fixed on, and you must gently direct it away from that side with your fingers when the pains commence, which thing will happen if you put your two fingers in front of it between its head and the side. It will slip off, but you must also not push it too much so as to keep it from getting fixed again on the other side.

Thirty-third question

JUSTINA: Can a child, lodged in this way by its head, suffer injury if it is not helped?

CHRISTINA: (1) If a child is too firmly fixed, it cannot get free without help until it dies, and mother and child will also probably remain to-

165. Siegemund refers here to the molding of the child's head that occurs as a result of prolonged compression during its passage through the birth canal and that enables that passage.

gether, for if it is to slip off on its own, it requires time and energy. (2) Also, if two or three days of labor ensue, even if it goes as best it could on its own, the child seldom survives when it lasts that long, as can be seen with the bad heads that newborn children sometimes bring with them. (3) In question number 30 [*sic*], you said children lodged on the sharebone could easily turn over because they lie up high in the hollow belly. If such turning over comes to pass, mother and child will be in danger of losing their lives. (4) Children whose heads are fixed in this way cannot only turn completely over, but they also tend to present with their little hands, if not with both, then usually with one of them over their head in the birth passage, which can happen as long as the waters are intact since they have room and space to double up and turn over.

Thirty-fourth question

JUSTINA: Can this fixing of the child's head come to pass in no other way?

CHRISTINA: As far as I know from your instruction and otherwise, I can only judge that the fixing of children's heads mostly results from the following causes, to wit, (1) with very stout bellies, where it is hard to prevent; (2) it also comes to pass with those bellies that hang too far out over the sharebone; and (3) this lodging and fixing happens when the woman has a lot of water.

Thirty-fifth question

JUSTINA: How long does a child remain lying in the right posture when it has turned before delivery?

CHRISTINA: Children lie turned for birth for different lengths of time, one longer than the other, depending on whether the child is large or small, whether the woman has a little or a lot of room in her belly, and whether the child lies in a large or small amount of water. If it is roomy and the child small, then they seldom remain lying in the right posture when the pains come, or even without pains while the woman is still on her feet. If only one thing is changed, either the child is large or the waters scant, they lie readily turned to the posture for birth during the final month and they are prevented from turning over because they have so little room. Thus in many bellies, where everything is tight, the child must stay turned for birth all the while, for eight or even ten weeks. And I have never seen a dearth of such mothers or children. With such tight bellies and with so little water a child cannot flip over. But when the child does not turn to the right posture in a timely fashion, it can thereafter, when it has already grown too large, hardly turn to the right posture for birth. From the cases I have described here, you can see that a child can turn in the belly sooner with one woman than with another.

Thirty-sixth question

JUSTINA: What is the cause when a child lies right for birth and yet cannot be born so mother and child sometimes lose their life?

CHRISTINA: There are various causes of a child's not being able to be born, even though it lies right for birth. These are to be noted, to wit, (1) If the child's head is too large, which question 31 already addressed somewhat; (2) if its shoulders are too broad; (3) if the child's head fixes on the woman's sharebone; (4) if the head lies crooked before the birth passage, and (5) if the head lies too far toward the rectum.

These five kinds of birth are generally called natural births, and they are too, as long as they are handled properly as you instructed me above concerning these cases—and still more—of children in the right posture.

Thirty-seventh question

JUSTINA: How are you supposed to manage or treat these births, inasmuch as they are called natural births and yet they are dangerous? How do you stave off danger?

CHRISTINA: A midwife must know how to distinguish these births by touching right at the start of labor, if misfortune or at the very least a difficult labor are not to ensue; they can thus (with God's blessing) be prevented. For one thing, if the child's head is too large, such children generally lie deep in the belly and because of their size cannot get as close to the entrance to the birth passage as others. A midwife must assist a child with the customary two fingers. She must span it, and she will thereby note how hard she should encourage the woman to push so as to avoid injury, since if the mother pushes too hard, mother and child run the risk of losing their lives. I speak not only of unnecessary use of medicaments or forcing the throes by means of medicaments, but also of assistance that is too forceful or urging the woman to push more than the child can give way. The midwife must get this by touching. For if the belly is naturally disposed to hard births and she pushes a child with a large head that needs a wide opening faster than the belly can give way, the child will get squeezed together because it cannot retreat and it will more likely suffocate than if you stick with the natural labor pains. The natural pains do not force the birth any harder than what it can take. So injury seldom occurs. In circumstances like these, the midwives and the women present generally pronounce such pains too weak, leading to error and danger for mother and child. If the case is to be handled right, then it is best for the midwife to assist simply by guiding the child with the two fingers she introduces and allowing herself to be guided by how much it gives way or can give way from pain to pain; she can urge the woman to push as necessary as long as she feels it is giving way properly. As soon, however, as she notices the woman is pushing harder than it is yielding, she must urge her to stop pushing. For if she pushes even a little too hard, the head, which otherwise must be pointy if it is to be a happy birth, will be flattened. I have always found this manner of assistance, in keeping with your teaching, to be beneficial with children whose heads were too large: the head can enter the birth passage stretched out as long as necessary, so long as it is not exaggerated. Children are sometimes thereby in danger of losing their lives if the head is too large and it takes too long for the head to become pointy, but likely the mother will not lose her life.

Second, as pertains to a child with very broad shoulders, such a birth must also be handled prudently like the aforementioned. You need to note this difference: the head is always easier to grasp because the shoulders free up the head so it does not lie wedged in so tight. And I have often found myself deceived in that I have thought because it had so much room, it was only a matter of the lack of labor pains, and so I have vigorously exhorted her to push. Then the head smashes through as a result of such force before the shoulders have room enough to follow and the child is thus in peril of its life, which does not happen if the child is pushed no harder than Nature itself pushes it forward. For as I have learned, God be praised, when Nature pushes it forward, it is wedged in tight and passes through at the same time. You have to be careful of this and gently govern and guide the child with your fingers. Once the child's head has been guided in, the child forces the mother's belly to open up from pain to pain. And the midwife must not stretch the neck with her hands or fingers. Above you showed me the wrongheadedness of this widespread notion, namely, this violent stretching hurts the woman's belly and causes swelling before the child comes forth and gets to this part, so the pain of breaking through is all the greater on account of the swelling and the injured belly, which I have perceived can cause more harm than good. Proper assistance tends to the child's head and where it is stuck tightest and does not occur in front in the neck where there is yet no child.

Thirty-eighth question

JUSTINA: What you say is right; the best assistance is with the child. But when children have bodies that are way too large, they can quite easily die during the delivery before you can get their shoulders loose, however adroit the midwife may be. What else can you tell me?

CHRISTINA: As far as the remaining three kinds are concerned, proper assistance always occurs in a timely fashion when you help the child's head into the birth passage; that is how you best prevent it from getting crooked or fixed. If no one assists, only a difficult birth can follow and probably the death of mother and child. But you should note and be familiar with the way in which the other kinds differ from the two deliveries described above, namely, in the third kind of the delivery,

When the child's head has fixed on the sharebone, the child does indeed lie deep inside, like children with large heads, but toward the rectum; the woman's belly is hollow and emptied of the child, whereas otherwise with large-headed children it is completely filled up. But if, as in the fourth kind, the child lies with its head crooked, you can see this by the child's ear: it can easily be reached with the customary two fingers (whereas with the right posture it lies so deep inside you cannot do this). Thus the opening in the child's head, namely, the fontanel, lies to one side and is hard to reach, whereas if the child lies right, you can and must readily reach it. And fifth, when the child runs or has run up sharply against the rectum, experience tells us that it also lies deep inside or high up, like children with big heads. Because it is headed toward the backbone with the front part of its head, it pushes toward the birth passage with the back of its head until the child's shoulders fix on the sharebone. Thereafter it remains fixed until it dies and the mother probably also loses her life along with it. There is no better way to recognize this (because children's heads are open in front as well as in back) than if you find the pains are for naught and pass in vain and the child does not budge in the least, which situation I have experienced to my detriment. So you must introduce the customary two fingers up by the sharebone and deep inside, and you can quickly find and feel the neck and the shoulder, which cannot be reached when children lie right. I have always allowed myself to be guided by this and will do so in future.

Thirty-ninth question

JUSTINA: Can a child turn in and out of position for birth as it pleases?

CHRISTINA: A child can turn in and out of position in one belly sooner than in another. And in the case of bellies where children can change position and turn over, women must expect all kinds of deliveries. The child's position changes with each pain, so when the waters break, the birth will take place as the child lies, however it may lie. Thus if the child is to be saved, a midwife must be guided by the way it lies when the waters are still intact.

Fortieth question

JUSTINA: What kind of deliveries ensue if a child turns to a wrong posture?

CHRISTINA: If it is a matter of bellies that naturally have difficult labor, the child will very likely die if it turns to the wrong posture. If it is a matter of bellies that deliver easily, the difficulty will not be too great and the child will be born feet first, though sometimes the child is endangered. So it is best if it can be prevented, insofar as it is possible.

Forty-first question

JUSTINA: Does no child present unnaturally other than those who turn like that?

CHRISTINA: There are to be sure unnatural births that are not possible to prevent, as, for example, when the child never turns to the right posture. In this case, the midwife is not responsible for the unnatural birth because it is impossible to prevent and it must ensue as God wills it, if the midwife but see to it that she want not for diligence. It is indeed dangerous for the child but not for the mother, as long as a sensible midwife is attending.

Forty-second question

JUSTINA: What if a child presents with its little hand? Must mother and child perish?

CHRISTINA: There are various kinds of arm presentations, and the mothers can be saved with the help of God in all circumstances as long as an experienced midwife is present. But speaking of the child, naturally you are better able to save it with one kind of a presentation than with another.

Forty-third question

JUSTINA: In how many different manners or ways can children come forth hands first, and what is the difference in these?

CHRISTINA: You showed me seven such arm presentations in your lesson above—at the moment I am familiar with only five of them, namely, (1) When the child tries to come forth head first with one little hand, (2) when the child raises both hands above its head or threatens to do so, (3) when the child lies on its back with its arm stretched out behind it, (4) when the child comes forth by its rump, but sometimes to one side with a little arm coming forth in front of it, and (5) when, as with some women, the child comes before or into the birth passage with a hand and foot.

The first presentation, when the child presents with its head and one little arm, is dangerous if it is not prevented by pinching or bringing back the arm, as you instructed above. The second presentation is exceedingly bad: the child presents with both little hands in front of the head and pinching cannot aid it; otherwise it will break its neck, and there is nothing else to do but undertake the difficult turn, which you have described at length. With the third presentation, when the child lies on its back with its arm stretched out below, you can snuggle your hand through beneath the little arm. But you have to hold onto the arm a little with your external hand (but not too much, only as necessary) in order to get by it, which thing the examination will indicate. This delivery is hard, but possible to manage, and the sooner the better. And it is also possible when the waters are still intact, if you are careful, to prevent the child from coming forth that way at all. The fourth presentation is when the child presents with its rump and a little hand. This is not always dangerous, even if the midwife does not know how to intervene; I had a happy outcome even before I knew much about how to assist. But you cannot expect it always to go well, and if it becomes dangerous, then it presents the greatest danger to the child. For this reason it is necessary to help the child through delivery quickly. A midwife can help out better here than with the previous delivery because the rump is not as round and it is not as dangerous to take hold of it as the head. The arm goes back on its own when you help the rump into position. For when the child lies straight, it has the power up top to pull back its arm on its own. With the fifth presentation, when the hands and feet lie beneath the child and one arm pushes forward, I have found the following to work, though I did not know what I was doing when I did it: if I simply put the arm right back up inside the neck, the feet are forced to come out in front, especially when the little arm is held back. And sometimes pushing it back like that helps so neither mother nor child is endangered, but it is an unnatural birth. And if danger ensues, then it is greatest for the child.

JUSTINA: I am pleased that you have pretty well grasped my instruction and view. Go on.

CHRISTINA: With regard to a foot and hand presenting at the same time, it is one thing as long as the feet are kept together. Although only one of them presents, you need to search for the other one immediately, for they are seldom far apart, and the hand will not present an obstacle. For if you can get the feet to come forth together, the hand will withdraw on its own and nothing will obstruct the delivery, as with the rump and hand, as long as the child is guided to the right posture for birth (I call it turned around right), as well as it can be, even though it has to be born unnaturally. If, however, it is not properly guided into the birth passage, then such a delivery can become quite dangerous for mother and child.

Forty-fourth question

JUSTINA: Are the arm presentations you have talked about easier or more difficult to manage than when a child presents with its feet alone or with the rump alone, without the arm, and which among these is the hardest and most difficult birth?

CHRISTINA: These arm presentations vary. In one way they go worse and in the other better, as long as they are attended to in a timely fashion with good sense and knowledge, for delivery by the feet—whether the child presents feet first or has to be turned by the feet—is always dangerous, inasmuch as unnatural births are in and of themselves very difficult and dangerous for the child. Though deliveries with the rump first should not be seen as easy, it is nevertheless true, as you instructed me, that deliveries with the feet first endanger the child's life more than those with the rump first. But delivery by the rump is harder for the mother than delivery by the feet because the child comes forth doubled up. But the woman need fear no harm from it, and the child is not readily imperiled.

Forty-fifth question

JUSTINA: For what reasons does one child sooner survive than another, and what advantage does a child have if it is born by the rump over a child that comes forth feet first?

CHRISTINA: One woman's belly is better suited for giving birth than another's on account of the hard closure of the inward mouth. A delivery with the feet first cannot force open such a hard closure as can the rump; the latter pushes the child down with greater force, and the inward mouth gives way sooner from the violent throes and the doubled-up child than when the child comes with its feet out in front. And precisely for this reason I consider birth with the rump first to be safer than birth with the feet first, inasmuch as the doubled-up child forces the mouth of the womb to open up sufficiently so the child can pass through it. If you are careful, it will not harm either the child or the mother.

Forty-sixth question

JUSTINA: Are there any bad births, besides the ones you have recounted, when no part of the child comes forth?

CHRISTINA: There are two types of births when nothing presents. The one is when the child lies on its back athwart the womb so nothing is visible before or in the birth passage. The other is when the child lies athwart the womb on its side. You cannot see anything coming forth, and the entrance to the birth passage is completely plugged up with its soft side. Neither can be born unless they be turned.

Forty-seventh question

JUSTINA: How do you distinguish these presentations from one another?

CHRISTINA: You can recognize the first presentation because you cannot reach it with your fingers because the child's back cannot give way until it is broken. So you need to employ your entire hand as soon as it is possible to get it through the opening in the inward mouth. Then you can feel the cartilage of the backbone. By contrast, in the other sort of presentation, you can reach the child with your fingers because the child's side will give way and the child feels soft and slippery when there are no pains.

Forty-eighth question

JUSTINA: How can you assist with these two presentations?

CHRISTINA: The first presentation is harder to recognize than the second one because you cannot reach it. According to your teaching, you can assist in this case in no other way but by searching for the feet and turning the child; the sooner this comes to pass, the better. The second turn is, however, more difficult than the one with the back I have just addressed. Because the child is jammed in tight by its side, you have to push it away and back. Thereafter you turn it like the other one; you have to seek the feet and take hold of them.

Forty-ninth question

JUSTINA: Are those bad postures the infants assume just as bad when the waters are still intact?

CHRISTINA: There are some postures children assume that are just as bad with the waters still intact, like the one with the feet first, the one where the navel string comes forth in front of the head, and the one where the hands and head emerge together. In addition, there is the one when hands and feet present together. Likewise, there are the cases when the child lies with its knees facing the birth passage and when the child presents with its shoulder or lies athwart the womb on its back or presents with its backside, as well as the case of twins. In all these deliveries the children tend to assume these postures with the waters still intact. However, the ones when the child presents with one or both little hands, with its stomach or its breast, or with its shoulder or its side, these are all postures that are forced when the waters break and the children are dry and the violence of the throes pushes the children so inefficiently. But a midwife with the sound knowledge that comes from experience can prevent these postures when the waters are still intact or as soon as they break as long as the mouth of the womb is open.

Fiftieth question

JUSTINA: Is it possible to turn children in these dangerous postures, can they survive, and can their mothers be saved? Is it best to turn the child when the waters are still intact or when they break?

CHRISTINA: It is always possible to turn children in those postures and at the very least to save the mother, and this is best done when the waters are still intact, as soon as the belly has opened up. Children can be more readily turned for a natural birth[166] when the waters are intact than after they have broken and it is no longer possible to bring the child's head into position for a proper birth; rather, in this case you must simply take hold of the feet. It is easier to manage the child that way than by the head because the woman's belly no longer has the room when the waters have run off that it has when the waters are still unbroken. If, however, the skin around the waters is so delicate that you have to fear it will tear when you touch her, it is better for you to wait until the waters break on their own. For as long as the waters are intact, the child can still turn, as God decrees it, sometimes

166. That is, a birth head first.

for the worse, sometimes for the better. Whatever happens, the midwife need not search her conscience as she does when the delicate skin tears while she is touching her. I have found, however, as you have also taught me, that it seldom comes to pass that the skin tears as a result of touching as long as you are careful of it and do not pinch it or tear at it with sharp fingernails. If the child cannot be aided with the waters unbroken, the midwife must introduce her hand when the waters break and undertake to turn the child, either by the feet or the head. You can do this most easily before the waters have completely run off, for the more you allow the waters to run off, the harder turning is on mother and child. The hand you have introduced can hold back the water until you have gotten hold of the child, as long as the midwife knows how to watch out for it. When the child has been gotten hold of, the waters your hand has held back will help it and I let go of them and turn the child very gently, which is a big help, so mother and child suffer no injury, especially if the woman is inclined to give birth easily, as you reported above. If, however, she is inclined to give birth with difficulty and the child has to be seized by the feet, it will be impossible to help the child, but it is certainly possible to help the mother, and at the very least you will spare her several days of hard labor.

Fifty-first question

JUSTINA: What is to be done if the waters break before the midwife is on hand?

CHRISTINA: If the waters break before the midwife is there and the child lies right and the pains and the proper opening are present, the child will not wait for the midwife; rather it will be born happily without her. If, however, the child lies wrong and the mouth is open, even though there are no pains, it is then necessary to turn it soon, for the longer it lies in the wrong posture, the more it sinks down with the waters that are trickling out toward the exit and it is all the harder to turn it thereafter and more dangerous for mother and child. If pains do not set in soon after turning the child, then there is, as far as I know, nothing that can be done and you have to wait for them. And this is true, as you yourself know and taught me above. It, however, seldom comes to pass that the labor pains do not set in after you turn the child; if, however, they do not ensue, you cannot readily force them. Turning the child helps so much that, with God's help, you can save the woman. If there are labor pains present, then it is all the more necessary to hurry up and turn the child so it cannot get wedged in too hard in the wrong posture.

Fifty-second question

JUSTINA: Waters often break several days before the delivery. What is to be done in this case?

CHRISTINA: If there are no labor pains and the child lies right, nothing need be done but wait for the pains and the appointed hour of birth, and so it will go well for mother and child.

Fifty-third question

JUSTINA: Often there are sufficient labor pains but the delivery nevertheless stretches over several days and the child is stillborn as a result. Did someone make a mistake?

CHRISTINA: There are a lot of mistakes that can be made in this case since the matter is hidden and does not meet the eye. It is certainly true that you have to await the hour of birth when there is nothing wanting. But where something is wanting, the appointed hour of birth can also be missed, as when the child lies wrong, as you recounted above. A lot of mistakes are made in that case. Indeed, children can lie right, but get lodged crooked on one side or the other or on the woman's sharebone, and they do not get free until after they have died. When labor starts, these children usually lie high up inside. Thus a midwife can make a mistake because she does not get at the child until it is dead and slips off and pushes in closer to the birth passage.

Fifty-fourth question

JUSTINA: Can you always reach the child when labor starts, however it lies, whether or not the waters have broken?

CHRISTINA: You can and must be able to reach the child when labor starts so if there is something there that is not good for the delivery, nothing bad will happen.

Fifty-fifth question

JUSTINA: What births can Nature force on its own?

CHRISTINA: When the child enters the birth passage with its head in the right posture, it can easily force the birth.

Fifty-sixth question

JUSTINA: What births can Nature force by itself with strong labor pains, when the child lies wrong, especially in the case of a strong woman, so both mother and child sometimes come away with their lives?

CHRISTINA: These births vary, to wit, those when the feet present; the child may turn its face toward the mother's back or toward her belly. It will work if its face is toward the belly, though with greater difficulty because it is easy for it to get stuck by the chin on the mother's sharebone. The child will sooner suffocate in this case than with the other presentation when its face faces her back. I have seen even more difficult cases, as when the child presents with its face; though it is considered to belong to the impossible deliveries, the powerful throes nevertheless force it, if the woman is strong, and it ends happily. In addition, if the hands and feet present simultaneously, the pains mostly force the feet because they can slide the most. For the child tries to come forward with its feet when the pains come because it has more vim in its feet than its hands for sliding, whereby the delivery can finally take place. Then too, if it presents with its knees, Nature forces it, just as with the feet. Furthermore, if it presents with its backside, sometimes an arm comes along with the rump and still it is not dangerous. Finally, it may present with its arm, as reported above, because its feet stay back a little ways; nevertheless, this too is not dangerous.

Fifty-seventh question

JUSTINA: Is it necessary, when the pains are weak, to strengthen them?

CHRISTINA: Sometimes, when the pains are too weak, it is necessary to strengthen them, as long as you do it at the right time. But it is often highly injurious.

Fifty-eighth question

JUSTINA: When is it necessary to strengthen the pains or to push, and when is it dangerous?

CHRISTINA: It is necessary and responsible to push when the child lies in the birth postures, described above, since Nature can force them. If, however, the child lies in a posture that is not good, as reported already, it is highly injurious to force the pains, for it will cost mother and child their lives.

Fifty-ninth question

JUSTINA: But if the child lay in a very bad posture for birth so it were impossible to deliver it, should you not give physics to strengthen the pains? And why not?

CHRISTINA: You must not give physics to strengthen the pains because a child lying in a bad posture cannot be born in that manner and the pains squeeze it so tight that turning it thereafter is all the harder. Such unnecessary pushing will sap the laboring woman of all her strength, whereas otherwise, when there is no pushing, she can hold out longer and all the better endure having the child turned, if it is possible to get someone to turn it.

Sixtieth question

JUSTINA: How is it best to undertake turning the child? With the woman standing, sitting, or lying down?

CHRISTINA: It is best to undertake turning the child with the woman lying down because the child can be gotten back that way. The belly is roomier when the woman is lying down than when she is standing or sitting; thus the turn is easier for mother and child.

Sixty-first question

JUSTINA: What is best for a woman in labor? Walking or standing?

CHRISTINA: Walking and standing at the onset of labor is not to be repudiated as long as the inward mouth does not or cannot open up. As soon as there is an opening, walking is harmful, because the fetus cannot enter the birth passage on account of the mother's walking, especially if the child is good sized. When it finally has to penetrate and does penetrate from the force of the throes, violence is done to the child while the mother is walking. It could certainly enter the birth passage while she is standing, but the woman will have a very hard time if it is delayed a little. This no doubt occurs with rapid births, but very rarely.

Sixty-second question

JUSTINA: People say the afterbirth sometimes precedes the child. Is that true then? How does it happen, and how do you know or notice it when it is present?

CHRISTINA: It indeed sometimes comes to pass that the afterbirth precedes the child and obstructs the delivery. I do not know how it happens, but I know full well that it does. A proper midwife can readily recognize it by touching if she but introduce her first two fingers into the inward mouth. In this state of things, you cannot feel the sac, which you otherwise can and must always feel, but instead when you touch her, you feel a piece of flesh like a liver, so that is why they call it a liver cake in German. Thus you also cannot feel through the thick flesh how the child lies the way you otherwise can always feel it during labor. For this reason it is easy for a midwife to discern. There is also always a great flood of blood before and during the delivery so mother and child are gravely endangered if they are not properly assisted. And even if children lie right for birth, most of them perish if they are not aided promptly—not to mention when the children are in the wrong posture, when the danger is even greater.

Sixty-third question

JUSTINA: How do you intervene in cases when the afterbirth precedes the child so mother and child need not perish?

CHRISTINA: I know of no way to intervene when the afterbirth precedes the child but—as I have done in such cases in accordance with your teaching—to employ a knitting needle or hairpin or similar tool to pierce the thick flesh, that is, the so-called liver cake. You have to do this carefully so as not to reach the child. I have laid the knitting needle or hairpin, when there was no instrument at hand made for that purpose, on top of the two fingers I introduced into the woman's birth passage, guided it in and touched the liver cake gently and carefully on account of the point and quite gently pierced the liver cake and then stuck one finger in after it once it was pierced. The waters will come forth as usual then, and the flood of blood will soon cease. The waters, however, help to make the little hole bigger so I can get in it with both my fingers. With these two fingers you can make the hole in the liver cake large enough for you to get through to the child. If the child lies right for birth then and is still alive, mother and child will be delivered happily. If it is already dead on account of the flooding that has just occurred, the mother can nevertheless be aided. If the child lies wrong, you must aid it as it lies by turning it, as you have described, to save the mother, so long as she was not aided too slowly when she flooded so she no longer has the strength to recover.

Sixty-fourth[167] *question*

JUSTINA: What is to blame for that prodigious flood of blood that often takes place before delivery and before there are labor pains and also for the heavy bleeding that takes place at the onset of labor so mother and child are in danger of losing their lives since the pains are thereby weakened and the mother and child sapped of all their strength? How do you help when no medicaments will take effect, as in cases like those I have often treated?

CHRISTINA: I do not know what is to blame or what the cause is for flooding like that. As far as I know, when there is prodigious flooding before delivery, it is always when the afterbirth is in front of the child. As soon as this happens, bleeding sets in and flooding follows, be it before the delivery or when labor starts. It cannot be readily quelled with physics, except with the method previously described. I have variously seen the situation become very dangerous before they would accept this method of piercing the afterbirth, when in fact it mostly works well. At the very least the mother is saved, often the child as well, if you do it in plenty of time. I have also seen both of them perish in these situations when they did not submit to this intervention and when at that same time I still did not know how to do it.

Sixty-fifth question

JUSTINA: In what kind of delivery or posture can the navel string slip out in front of the child or come forth first?

CHRISTINA: The navel string can slip out in front and enter the birth passage first in all deliveries and postures of children. Especially if it is long and the child has a lot of room in the waters, it easily slips through.

Sixty-sixth question

JUSTINA: What kind of danger does it present for mother and child when the navel string comes forth first?

CHRISTINA: It is always dangerous for the child when the navel string comes forth first, but not very easy for the mother either. If the child lies right for birth and the navel string has slipped out past its head, it is not dangerous for the mother, but certainly for the child if it is not put back soon. If it is put back, which often happens, the child will survive because it is in the right posture for birth. But in other manners of difficult labor it gener-

167. In the original misnumbered as the fifty-forth question.

ally perishes, not only because of the navel string, but also on account of the unnatural birth, inasmuch as it is generally a sign of a weak child. And in such births the mother can be endangered as well, depending on how the child lies.

Sixty-seventh question

JUSTINA: Is one posture or another more dangerous for the child if the navel string comes forth first?

CHRISTINA: If the navel string is delivered first, the child is always in peril of its life, though in one posture more than in another. For example, if the child lies right and the midwife but puts the navel string back inside and is able to hold it back, by whatever means possible, nothing will happen to mother or child. So when I have tried to get the navel string back behind the child's head and could not hold it with the customary two fingers, I have placed a fine soft scrap of cloth, as you taught me, between the head and the belly. It could thereby be held back, and a happy birth ensued for mother and child. If the child presents for birth with its rump and the navel string is present, you can preserve the child's life as in a natural birth when you know how to do that sort of thing. Putting the navel string back inside and putting a little cloth in front of it serves the purpose with all postures, where it can be put in, because the child can conserve its strength better and can endure longer in the case of a difficult birth. If the child's posture is very bad, it will have to die on account of the bad posture and unnatural birth, not on account of the navel string. But however strong the child may be, however right its posture may be, if the navel string is not kept inside, it will be hard to save the child's life, unless it be a woman who is disposed to quick and easy deliveries. This very seldom happens, however.

Sixty-eighth question

JUSTINA: Is it then not possible early on to prevent the navel string from slipping out first?

CHRISTINA: The navel string will not slide out until labor starts, when the child is pushed by the throes. Before this there is no distress, however the navel string be positioned, because it cannot be pinched in the roomy belly. It is therefore possible to keep it inside and prevent it from happening by breaking the waters as long as you do it early on in those deliveries where the child lies right. Because the navel string has not been out in front very long yet, it is all the easier to put it back inside. For the longer the waters remain unbroken—since they provide lots of room—the more and longer it slides around in the roomy water bladder, and the longer the child is in this state, the weaker it becomes in that the pains always pinch the navel string a great deal. Indeed, the child can also die before the waters break on their own. In fact, it is thereafter also very difficult to put it back inside when it has been out that far in that it is very slippery. You yourself know this better than I can answer you by referring to what you taught me.

Sixty-ninth question

JUSTINA: Can you always put the navel string back inside and keep it there?

CHRISTINA: You can very usefully put it back inside with some births, as has been reported. However, it is not possible to put it back inside in all cases.

Seventieth question

JUSTINA: In what postures can you best put the navel string back inside and keep it there, and in which of them is it not possible for you to keep it inside?

CHRISTINA: In most deliveries and postures you can put the navel string back inside soon after the waters break and you can keep it there by sticking in a soft little cloth, as I answered above in the sixty-seventh question[168] in accordance with your teaching, except where the child presents or threatens to present with its belly. In that case you cannot keep the navel string inside.

Seventy-first question

JUSTINA: Can you know before the waters break that the navel string is coming forth first? Is it necessary for a midwife to know this before the waters break?

CHRISTINA: Even if the waters are unbroken, it is quite easy to know and to feel whether the navel string is coming forth ahead of the child—and indeed with the first labor pains. The further along labor is, the better you

168. The original text incorrectly refers to the sixty-ninth question.

can discern it. It is also highly necessary for the midwife to know about this because the child's life is at stake, and it can be saved in the nick of time as long as it lies right for birth.

Seventy-second question

JUSTINA: Is it sometimes beneficial for the midwife to break the waters or not? Does it help or hurt?

CHRISTINA: It is highly necessary to break the waters in certain cases, and you can save the life of mother and child where Necessity demands it. As necessary as it is, however, it can also be injurious if it is done at the wrong time, and mother and child can be in peril of their lives, as you yourself know all too well.

Seventy-third question

JUSTINA: When is it necessary to break the waters and what are the benefits?

CHRISTINA: If the skin is too tough and the child is in the right posture for birth, you can break the waters. You can do the same with such bellies where it is possible for the child to turn during labor. This is easy to feel and to know. You can save the child's life by doing this as long as you do it in time and the child lies right for birth.

Seventy-fourth question

JUSTINA: Can an inexperienced midwife break the waters too early on before the hour of birth is at hand and therefore prematurely?

CHRISTINA: No midwife can break the waters prematurely, however clever or stupid she may be, for, as you instructed me above, the inward mouth does not give way before the appointed hour of birth and thus does not permit the waters to come forth into the straits[169] so a midwife who does not know what she is doing could break them. And one who does know what she is doing cannot do it at the proper time without an instrument. If she knows how to do it with an instrument, she knows she should not do it prematurely. You cannot undertake to break the waters without the woman's knowledge and consent; you cannot do it without gravely distressing and

169. In translating German *Dohne* as "straits," I take my cue from the Dutch translation of Siegemund's handbook. *Dohne* normally refers to a kind of bird trap.

endangering the child. Because it is unusual, the outcome would shortly punish an ignorant midwife, though an ignorant midwife can do no such thing [in the first place—*trans.*].

Seventy-fifth question

JUSTINA: You are right. That is all true. But what kind of danger does breaking the waters present if it is not done prematurely?

CHRISTINA: Breaking the waters is often dangerous—not on account of it being done prematurely but because of children lying wrong; when the waters are broken prematurely, a bad birth ensues, putting the child at risk of losing its life (which posture could still change during the birth before the waters break on their own). You spoke at length and in detail about this in the first part.

Seventy-sixth question

JUSTINA: How is it then with the delivery when there are twins or when there are more children present?

CHRISTINA: You can only assist twins and more infants in the postures they lie in, as with all the previously described turns in all sorts of births. However they lie, the first one has to come forth first, and the other gotten out afterwards. It would be the same if there were yet another one and even more. It is often easier to manage two or three small children than a right large one when it lies wrong.

Seventy-seventh question

JUSTINA: Do tell me whether twins always lie in one afterbirth[170] or whether each child has its own afterbirth, and if they share one, whether the waters break twice?

CHRISTINA: Where there are twins, the children do not lie in just one way with regard to the afterbirth. Sometimes each child has its own afterbirth; sometimes they lie together in one afterbirth, but there is a skin between them, and thus the waters break twice. If they lie in a single afterbirth with only a skin in between, then when the first is delivered, the other water bladder follows. If, however, each child has its own afterbirth, the afterbirth of the first child often comes forth before the second water bladder can

170. Siegemund conflates the amniotic sac and the placenta in the term "afterbirth."

enter the birth passage. And it has to be gotten out ahead; that generally happens if the navel string is short. Where, however, there is a long navel string, the first afterbirth often comes forth after the last child, especially when the second child is strong and can push out soon after the first one. Then the first afterbirth has to make way for the child and [the second child thus—*trans.*] pushes the first child's afterbirth back behind itself. That is why you have to be very careful to get out the afterbirth where there are two or more children, if slow, indeed unhappy, births are not to ensue, all of which I have observed in keeping with your manual and found to be so too.

Seventy-eighth question

JUSTINA: Does the afterbirth follow on its own, when there is only one child and things go naturally, or how do you help it out if it does not follow?

CHRISTINA: The afterbirth usually follows on its own, and it is easy to help it out by [having the woman—*trans.*] cough or sneeze as long as it does not fix on the inward mouth.

Seventy-ninth question

JUSTINA: If the afterbirth remains inside, is there no other cause but that it must be ingrown, and can you do something if it is ingrown?

CHRISTINA: The afterbirth is rarely ingrown, even though it may remain inside for several days. Many years can pass before you encounter an ingrown afterbirth, but you cannot entirely gainsay it, because it does happen. To my knowledge you cannot get it out if the afterbirth is ingrown; but when it is thought to be ingrown, when really it is not, many women lose their lives over it.

Eightieth question

JUSTINA: What is wrong then when the afterbirth remains inside and yet is not ingrown? How do you help it out?

CHRISTINA: If the afterbirth is disposed to remain inside and yet is not ingrown, you can easily help it out with a gentle intervention performed with sound knowledge: if you take hold of the navel string with your left hand and introduce the first two fingers of your right hand along the navel string, you will soon find where it is hung up, namely, behind the inward mouth. When you loosen and push it up with your two fingers, the afterbirth will easily come forth.

Eighty-first question

JUSTINA: You say in your answer to the seventy-ninth question that the afterbirth is seldom ingrown. How can that be? Must not the afterbirth always be ingrown because the child gets nourishment through the afterbirth and the navel string?

CHRISTINA: You know better than I can answer you that there are two ways in which the afterbirth can be attached, to wit, it can be attached so it does not come off during delivery or attached so it can detach itself in a natural way, inasmuch as it usually detaches itself during and after the birth and can easily be evacuated by coughing and sneezing, as has previously been sufficiently noted.

Eighty-second question

JUSTINA: Is there nothing you can do if the afterbirth is, as you think, quite firmly ingrown? Must all these women die?

CHRISTINA: I do not want to declare their lives forfeited. Nothing is impossible for God. But common sense tells us it is quite dangerous. Thank God these sorts of conditions seldom occur.

Eighty-third question

JUSTINA: Now I would like to hear again whether you applaud my advice against using various remedies with women in labor.

CHRISTINA: Dear sister! He who does not learn from another's misfortune can be neither advised nor helped. The story and the clear proof regarding Titia taught me so much that I will proceed cautiously my whole life long in such situations and will show the venerable doctors the obedience due them.

Eighty-fourth question

JUSTINA: Your resolve is proper and free of danger, but what do you think about tipping women? Can this tipping help or harm?

CHRISTINA: I have never witnessed this sort of tipping; I am thus of the opinion that it is irresponsible and contrary to all reason. Thank God I have acquired enough of a basis in the art as to how difficult births should and can be prevented in a timely fashion, however, only with divine intention and goodwill. Otherwise our helping hand is only a mortal work, all our knowledge too, only patchwork. But the help of the Lord blesses our profession and work.

Eighty-fifth question

JUSTINA: You certainly speak in a sufficiently Christian manner; nevertheless, I have been told you caused the death of sundry children because in unnatural presentations you delivered many a woman quickly and the child died as a result of this quick delivery. What then causes the child's death?

CHRISTINA: The primary cause is that I am not omnipotent like God; the secondary cause, however, is the child's lying wrong. Many a woman is not worthy of being delivered so quickly of a child that lies wrong, for she thanks neither God nor man, considers even less that life and death are in the hands of the Lord and that in such a dangerous situation He often helps through the good faith of a practiced midwife.

Eighty-sixth question

JUSTINA: Tell me as well how you understood my lesson on the birthing chair and what you believe to be the best place for the woman in labor—the birthing chair or the birthing bed.

CHRISTINA: During labor neither is to be despised, neither the chair nor the bed. Both are beneficial as long as the midwife is knowledgeable as to where the throes can best push the child forward and into the birth passage, which the midwife's touch must reveal. If, however, she does not understand how to do this, the birth can be delayed or become endangered elsewhere, for if it cannot enter the birth passage properly, it will necessarily be slowed down. If it is slowed down then, be she sitting or standing, and if this position is not changed, the child will be forced to go off crooked. Often children lie crooked from the beginning, so the midwife must not let herself get fixed on any other place but where the child enters or can enter the birth passage. Sometimes it comes to pass that the woman needs to lie on her side—contrary to all natural custom. So both are good, if you watch out for where it can or does work best. The birthing chair you showed me, which can be a comfortable bed at the same time, serves splendidly to prevent and aid any complications that occur.

Eighty-sixth [sic] question

JUSTINA: Is it necessary and responsible then in a difficult delivery when the child is dead and will not budge to put a crotchet in it and rescue the mother? Tell me your thoughts on the conclusion of my manual in the first part.

CHRISTINA: When the midwife is there right at the beginning of a difficult birth and knows how to assist by guiding the child's head, she should not, in my opinion, employ a crotchet. In unnatural deliveries, however, you can assist by turning the child, first of all because the child has plenty of room to turn as long as it has not entered the birth passage. If, however, the child gets stuck and has pushed too far into the birth passage (which never comes to pass as long as the child is alive), then it is indeed necessary to use a crotchet, for the child is dead, and the mother can thereby be saved. All this in keeping with your instruction above, whereby I got an idea from the copper engraving of the crotchets of what might be most serviceable in an emergency and impressed them upon my mind so as to make careful use of them in accordance with Divine Providence.

CONCLUSION

JUSTINA: Inasmuch as I am well pleased with your responses to my questions, I wish in closing that with assiduous practice you may become ever more secure in everything that this manual, which I wrote in good faith, aimed at and that you enjoy in your profession God's almighty support along with His rich blessing.

End of the second part.

Honor is God's alone.

APPENDIX A
TABLE OF CONTENTS AS IT APPEARED AT THE END OF THE GERMAN EDITION

First Index on the Chapters of the First Part, Which Consists of a Highly Necessary Instruction in Difficult Births and the Avoidance Thereof Where Possible, as well as Instruction in Deft Turning of Children Who Lie Wrong. Preliminary Account to the Well-Disposed Reader
[n.p.]

APPENDIX B
GLOSSARY OF NEW AND OLD GYNECOLOGICAL AND OBSTETRIC TERMS

afterbirth The placenta and fetal membranes expelled from the uterus after childbirth.
belly Abdomen, womb, birth passage.
cervical os/os Opening in the narrow outer end of the uterus.
crotchet Hook used in the seventeenth century to extract a dead fetus from the uterus.
dilatation of the cervix Opening of the cervix caused by contractions.
eclampsia Pregnancy-induced hypertension, coma, and convulsions arising from conditions during pregnancy or immediately after childbirth.
effacement of the cervix Thinning of the cervix as a result of contractions in preparation for birth.
fallen matrix, fallen womb, fallen belly Prolapsed uterus.
flooding Bleeding.
fontanel One of two soft spots between the unfused sections of the fetal skull.
forceps Tong-like instrument with pointed ends; in the seventeenth century it was used to extract a dead fetus from the uterus; modern forceps for extracting a live fetus were not yet known in the German territories.
fundament Fundus, upper rounded extremity of the uterus.
introitus Vaginal opening.
lingering labor Prolonged labor.
liver cake/cake Placenta.
matrix Uterus.
mole, molar pregnancy Tumor in the uterus; a molar pregnancy results when a sperm penetrates an egg without chromosomes and genes.
moon calf Mole.
mouth/inward mouth Cervix.
neck Vagina; occasionally "neck" is used to refer to the cervix.
navel string Umbilical cord.
presentation/to present Referring to the part of the fetus that emerges first; in an arm presentation, for example, the arm leads.
prolapsed uterus Uterus that has dropped from its normal position in the pelvic cavity and descended into and sometimes outside of the vagina.
prolapsed vagina Vagina that has dropped from its normal position, sometimes emerging from the body.

secundines The placenta and fetal membranes expelled from the uterus after childbirth.

sharebone Pubic bone.

sheath Vagina.

skin Membrane.

speculum Vaginal speculum; device used to spread apart and prop open the vagina in order to view the cervix; also used to dilate the cervix; the cumbersome early modern speculum generally consisted of three large spike-like blades that were inserted into the vagina and then distended with the turn of a screw, a ratchet, or a lever (see James V. Ricci, *The Development of Gynaecological Surgery and Instruments* [San Francisco: Norman Publishing, 1990]).

sponge Placenta.

teeming, to teem Pregnant, be pregnant.

touch Palpation of the cervix in order to determine the degree to which the cervix is dilated and the position of the fetus.

water bladder Amniotic sac.

SERIES EDITORS' BIBLIOGRAPHY

PRIMARY SOURCES

Alberti, Leon Battista. *The Family in Renaissance Florence.* Trans. Renée Neu Watkins. Columbia, SC: University of South Carolina Press, 1969.

Arenal, Electa, and Stacey Schlau, eds. *Untold Sisters: Hispanic Nuns in Their Own Works.* Trans. Amanda Powell. Albuquerque, NM: University of New Mexico Press, 1989.

Astell, Mary. *The First English Feminist: Reflections on Marriage and Other Writings.* Ed. and intro. Bridget Hill. New York: St. Martin's Press, 1986.

Atherton, Margaret, ed. *Women Philosophers of the Early Modern Period.* Indianapolis, IN: Hackett Publishing Co., 1994.

Aughterson, Kate, ed. *Renaissance Woman: Constructions of Femininity in England: A Source Book.* London and New York: Routledge, 1995.

Barbaro, Francesco. *On Wifely Duties.* Trans. Benjamin Kohl. In *The Earthly Republic,* ed. Benjamin Kohl and R.G. Witt. Philadelphia: University of Pennsylvania Press, 1978, 179–228. Translation of the preface and book 2.

Behn, Aphra. *The Works of Aphra Behn.* 7 vols. Ed. Janet Todd. Columbus, OH: Ohio State University Press, 1992–96.

Boccaccio, Giovanni. *Famous Women.* Ed. and trans. Virginia Brown. I Tatti Renaissance Library. Cambridge, MA: Harvard University Press, 2001.

———. *Corbaccio or the Labyrinth of Love.* Trans. Anthony K. Cassell. Second revised edition. Binghamton, NY: Medieval and Renaissance Texts and Studies, 1993.

Brown, Sylvia. *Women's Writing in Stuart England: The Mother's Legacies of Dorothy Leigh, Elizabeth Joscelin and Elizabeth Richardson.* Thrupp, Stroud, Gloceter: Sutton, 1999.

Bruni, Leonardo. "On the Study of Literature (1405) to Lady Battista Malatesta of Moltefeltro." In *The Humanism of Leonardo Bruni: Selected Texts,* trans. and intro. Gordon Griffiths, James Hankins, and David Thompson. Binghamton, NY: Medieval and Renaissance Texts and Studies, 1987, 240–51.

Castiglione, Baldassare. *The Book of the Courtier.* Trans. George Bull. New York: Penguin, 1967.

Christine de Pizan. *The Book of the City of Ladies.* Trans. Earl Jeffrey Richards. Foreward Marina Warner. New York: Persea Books, 1982.

———. *The Treasure of the City of Ladies.* Trans. Sarah Lawson. New York: Viking Penguin, 1985.

Clarke, Danielle, ed. *Isabella Whitney, Mary Sidney and Aemilia Lanyer: Renaissance Women Poets.* New York: Penguin Books, 2000.

Crawford, Patricia, and Laura Gowing, eds. *Women's Worlds in Seventeenth-Century England: A Source Book.* London and New York: Routledge, 2000.

Daybell, James, ed. *Early Modern Women's Letter Writing, 1450–1700.* Houndmills, England and New York: Palgrave, 2001.

Elizabeth I: Collected Works. Ed. Leah S. Marcus, Janel Mueller, and Mary Beth Rose. Chicago: University of Chicago Press, 2000.

Elyot, Thomas. *Defence of Good Women: The Feminist Controversy of the Renaissance.* Facsimile Reproductions. Ed. Diane Bornstein. New York: Delmar, 1980.

Erasmus, Desiderius. *Erasmus on Women.* Ed. Erika Rummel. Toronto: University of Toronto Press, 1996.

Female and Male Voices in Early Modern England: An Anthology of Renaissance Writing. Ed. Betty S. Travitsky and Anne Lake Prescott. New York: Columbia University Press, 2000.

Ferguson, Moira, ed. *First Feminists: British Women Writers, 1578–1799.* Bloomington, IN: Indiana University Press, 1985.

Galilei, Maria Celeste. *Sister Maria Celeste's Letters to Her father, Galileo.* Ed. and trans. Rinaldina Russell. Lincoln, NE, and New York: Writers Club Press of Universe.com, 2000.

Gethner, Perry, ed. *The Lunatic Lover and Other Plays by French Women of the Seventeenth and Eighteenth Centuries.* Portsmouth, NH: Heinemann, 1994.

Glückel of Hameln. *The Memoirs of Glückel of Hameln.* Trans. Marvin Lowenthal. New intro. by Robert Rosen. New York: Schocken Books, 1977.

Henderson, Katherine Usher, and Barbara F. McManus, eds. *Half Humankind: Contexts and Texts of the Controversy about Women in England, 1540–1640.* Urbana, IL: Indiana University Press, 1985.

Hoby, Margaret. *The Private Life of an Elizabethan Lady: The Diary of Lady Margaret Hoby 1599–1605.* Phoenix Mill, UK: Sutton Publishing, 1998.

Humanist Educational Treatises. Ed. and trans. Craig W. Kallendorf. I Tatti Renaissance Library. Cambridge, MA: Harvard University Press, 2002.

Joscelin, Elizabeth. *The Mothers Legacy to Her Unborn Childe.* Ed. Jean leDrew Metcalfe. Toronto: University of Toronto Press, 2000.

Kaminsky, Amy Katz, ed. *Water Lilies, Flores del agua: An Anthology of Spanish Women Writers from the Fifteenth through the Nineteenth Century.* Minneapolis: University of Minnesota Press, 1996.

Kempe, Margery. *The Book of Margery Kempe.* Trans. and ed. Lynn Staley. Norton Critical Edition. New York: W.W. Norton, 2001.

King, Margaret L., and Albert Rabil, Jr., eds. *Her Immaculate Hand: Selected Works by and about the Women Humanists of Quattrocento Italy.* Binghamton, NY: Medieval and Renaissance Texts and Studies, 1983. Second revised paperback edition, 1991.

Klein, Joan Larsen, ed. *Daughters, Wives, and Widows: Writings by Men about Women and Marriage in England, 1500–1640.* Urbana, IL: University of Illinois Press, 1992.

Knox, John. *The Political Writings of John Knox: The First Blast of the Trumpet against the Monstrous Regiment of Women and Other Selected Works.* Ed. Marvin A. Breslow. Washington, DC: Folger Shakespeare Library, 1985.

Kors, Alan C., and Edward Peters, eds. *Witchcraft in Europe, 400–1700: A Documentary History.* Philadelphia: University of Pennsylvania Press, 2000.

Krämer, Heinrich, and Jacob Sprenger. *Malleus Maleficarum* (ca. 1487). Trans. Montague Summers. London: Pushkin Press, 1928. Reprint, New York: Dover, 1971.

Larsen, Anne R., and Colette H. Winn, eds. *Writings by Pre-Revolutionary French Women: From Marie de France to Elizabeth Vigée-Le Brun.* New York and London: Garland Publishing Co., 2000.

de Lorris, William, and Jean de Meun. *The Romance of the Rose.* Trans. Charles Dahlbert. Princeton: Princeton University Press, 1971. Reprint, University Press of New England, 1983.

Marguerite d'Angoulême, Queen of Navarre. *The Heptameron.* Trans. P. A. Chilton. New York: Viking Penguin, 1984.

Mary of Agreda. *The Divine Life of the Most Holy Virgin.* Abridgment of *The Mystical City of God.* Abr. Fr. Bonaventure Amedeo de Caesarea, M.C. Trans. from French by Abbé Joseph A. Boullan. Rockford, IL: Tan Books, 1997.

Myers, Kathleen A., and Amanda Powell, eds. *A Wild Country out in the Garden: The Spiritual Journals of a Colonial Mexican Nun.* Bloomington, IN: Indiana University Press, 1999.

Russell, Rinaldina, ed. *Sister Maria Celeste's Letters to Her Father, Galileo.* San Jose and New York: Writers Club Press, 2000.

Teresa of Avila, Saint. *The Life of Saint Teresa of Avila by Herself.* Trans. J. M. Cohen. New York: Viking Penguin, 1957.

Weyer, Johann. *Witches, Devils, and Doctors in the Renaissance: Johann Weyer, De praestigiis daemonum.* Ed. George Mora with Benjamin G. Kohl, Erik Midelfort, and Helen Bacon. Trans. John Shea. Binghamton, NY: Medieval and Renaissance Texts and Studies, 1991.

Wilson, Katharina M., ed. *Medieval Women Writers.* Athens, GA: University of Georgia Press, 1984.

———, ed. *Women Writers of the Renaissance and Reformation.* Athens, GA: University of Georgia Press, 1987.

Wilson, Katharina M., and Frank J. Warnke, eds. *Women Writers of the Seventeenth Century.* Athens, GA: University of Georgia Press, 1989.

Wollstonecraft, Mary. *A Vindication of the Rights of Men and a Vindication of the Rights of Women.* Ed. Sylvana Tomaselli. Cambridge: Cambridge University Press, 1995. Also *The Vindications of the Rights of Men, The Rights of Women.* Ed. D. L. Macdonald and Kathleen Scherf. Peterborough, Ontario: Broadview Press, 1997.

Women Critics, 1660–1820: An Anthology. Ed. the Folger Collective on Early Women Critics. Bloomington, IN: Indiana University Press, 1995.

Women Writers in English 1350–1850: 15 published through 1999 (projected 30-volume series suspended). Oxford University Press.

Wroth, Lady Mary. *The Countess of Montgomery's Urania.* 2 parts. Ed. Josephine A. Roberts. Tempe, AZ: Medieval and Renaissance Texts and Studies, 1995, 1999.

———. *Lady Mary Wroth's "Love's Victory": The Penshurst Manuscript.* Ed. Michael G. Brennan. London: Roxburghe Club, 1988.

———. *The Poems of Lady Mary Wroth.* Ed. Josephine A. Roberts. Baton Rouge, LA: Louisiana State University Press, 1983.

de Zayas Maria. *The Disenchantments of Love.* Trans. H. Patsy Boyer. Albany, NY: State University of New York Press, 1997.

———. *The Enchantments of Love: Amorous and Exemplary Novels.* Trans. H. Patsy Boyer. Berkeley, CA: University of California Press, 1990.

SECONDARY SOURCES

Ahlgren, Gillian. *Teresa of Avila and the Politics of Sanctity.* Ithaca, NY: Cornell University Press, 1996.

Akkerman, Tjitske, and Siep Sturman, eds. *Feminist Thought in European History, 1400–2000.* London and New York: Routledge, 1997.

Allen, Sister Prudence, R.S.M. *The Concept of Woman: The Aristotelian Revolution, 750 B.C.–A.D. 1250.* Grand Rapids, MI: William B. Eerdmans Publishing Company, 1997.

———. *The Concept of Woman: Volume II: The Early Humanist Reformation, 1250–1500.* Grand Rapids, MI: William B. Eerdmans Publishing Company, 2002.

Andreadis, Harriette. *Sappho in Early Modern England: Female Same-Sex Literary Erotics, 1550–1714.* Chicago: University of Chicago Press, 2001.

Armon, Shifra. *Picking Wedlock: Women and the Courtship Novel in Spain.* New York: Rowman and Littlefield Publishers, Inc., 2002.

Backer, Anne Liot. *Precious Women.* New York: Basic Books, 1974.

Ballaster, Ros. *Seductive Forms.* New York: Oxford University Press, 1992.

Barash, Carol. *English Women's Poetry, 1649–1714: Politics, Community, and Linguistic Authority.* New York and Oxford: Oxford University Press, 1996.

Battigelli, Anna. *Margaret Cavendish and the Exiles of the Mind.* Lexington, KY: University of Kentucky Press, 1998.

Beasley, Faith. *Revising Memory: Women's Fiction and Memoirs in Seventeenth-Century France.* New Brunswick: Rutgers University Press, 1990.

Beilin, Elaine V. *Redeeming Eve: Women Writers of the English Renaissance.* Princeton: Princeton University Press, 1987.

Benson, Pamela Joseph. *The Invention of Renaissance Woman: The Challenge of Female Independence in the Literature and Thought of Italy and England.* University Park, PA: Pennsylvania State University Press, 1992.

Benson, Pamela Joseph, and Victoria Kirkham, eds. *Strong Voices, Weak History? Medieval and Renaissance Women in Their Literary Canons: England, France, Italy.* Ann Arbor: University of Michigan Press, 2003.

Bilinkoff, Jodi. *The Avila of Saint Teresa: Religious Reform in a Sixteenth-Century City.* Ithaca, NY: Cornell University Press, 1989.

Bissell, R. Ward. *Artemisia Gentileschi and the Authority of Art.* University Park, PA: Pennsylvania State University Press, 2000.

Blain, Virginia, Isobel Grundy, and Patricia Clements, eds. *The Feminist Companion to Literature in English: Women Writers from the Middle Ages to the Present.* New Haven, CT: Yale University Press, 1990.

Bloch, R. Howard. *Medieval Misogyny and the Invention of Western Romantic Love.* Chicago: University of Chicago Press, 1991.

Bornstein, Daniel, and Roberto Rusconi, eds. *Women and Religion in Medieval and Re-*

naissance Italy. Trans. Margery J. Schneider. Chicago: University of Chicago Press, 1996.

Brant, Clare, and Diane Purkiss, eds. *Women, Texts and Histories, 1575–1760*. London and New York: Routledge, 1992.

Briggs, Robin. *Witches and Neighbours: The Social and Cultural Context of European Witchcraft*. New York: HarperCollins, 1995; Viking Penguin, 1996.

Brink, Jean R., ed. *Female Scholars: A Tradition of Learned Women before 1800*. Montréal: Eden Press Women's Publications, 1980.

Brown, Judith C. *Immodest Acts: The Life of a Lesbian Nun in Renaissance Italy*. New York: Oxford University Press, 1986.

Brown, Judith C., and Robert C. Davis, eds. *Gender and Society in Renaissance Italy*. London: Addison Wesley Longman, 1998.

Bynum, Carolyn Walker. *Fragmentation and Redemption: Essays on Gender and the Human Body in Medieval Religion*. New York: Zone Books, 1992.

———. *Holy Feast and Holy Fast: The Religious Significance of Food to Medieval Women*. Berkeley: University of California Press, 1987.

Cambridge Guide to Women's Writing in English. Ed. Lorna Sage. Cambridge: Cambridge University Press, 1999.

Cavanagh, Sheila T. *Cherished Torment: The Emotional Geography of Lady Mary Wroth's "Urania."* Pittsburgh: Duquesne University Press, 2001.

Cerasano, S. P., and Marion Wynne-Davies, eds. *Readings in Renaissance Women's Drama: Criticism, History, and Performance, 1594–1998*. London and New York: Routledge, 1998.

Cervigni, Dino S., ed. *Women Mystic Writers. Annali d'Italianistica* 13 (1995) (entire issue).

Cervigni, Dino S., and Rebecca West, eds. *Women's Voices in Italian Literature. Annali d'Italianistica* 7 (1989) (entire issue).

Charlton, Kenneth. *Women, Religion and Education in Early Modern England*. London and New York: Routledge, 1999.

Chojnacka, Monica. *Working Women in Early Modern Venice*. Baltimore: Johns Hopkins University Press, 2001.

Chojnacki, Stanley. *Women and Men in Renaissance Venice: Twelve Essays on Patrician Society*. Baltimore: Johns Hopkins University Press, 2000.

Cholakian, Patricia Francis. *Rape and Writing in the Heptameron of Marguerite de Navarre*. Carbondale and Edwardsville, IL: Southern Illinois University Press, 1991.

———. *Women and the Politics of Self-Representation in Seventeenth-Century France*. Newark: University of Delaware Press, 2000.

Christine de Pizan: A Casebook. Edited by Barbara K. Altmann and Deborah L. McGrady. New York: Routledge, 2003.

Clogan, Paul Maruice, ed. *Medievali et Humanistica: Literacy and the Lay Reader*. Lanham, MD: Rowman and Littlefield, 2000.

Clubb, Louise George. *Italian Drama in Shakespeare's Time*. New Haven, CT: Yale University Press, 1989.

Conley, John J., S.J. *The Suspicion of Virtue: Women Philosophers in Neoclassical France*. Ithaca, NY: Cornell University Press, 2002.

Crabb, Ann. *The Strozzi of Florence: Widowhood and Family Solidarity in the Renaissance*. Ann Arbor: University of Michigan Press, 2000.

Cruz, Anne J., and Mary Elizabeth Perry, eds. *Culture and Control in Counter-Reformation Spain.* Minneapolis: University of Minnesota Press, 1992.

Davis, Natalie Zemon. *Society and Culture in Early Modern France.* Stanford: Stanford University Press, 1975, esp. chaps. 3 and 5.

———. *Women on the Margins: Three Seventeenth-Century Lives.* Cambridge, MA: Harvard University Press, 1995.

DeJean, Joan. *Ancients against Moderns: Culture Wars and the Making of a Fin de Siècle.* Chicago: University of Chicago Press, 1997.

———. *Fictions of Sappho, 1546–1937.* Chicago: University of Chicago Press, 1989.

———. *The Reinvention of Obscenity: Sex, Lies, and Tabloids in Early Modern France.* Chicago: University of Chicago Press, 2002.

———. *Tender Geographies: Women and the Origins of the Novel in France.* New York: Columbia University Press, 1991.

———. *The Reinvention of Obscenity: Sex, Lies, and Tabloids in Early Modern France.* Chicago: University of Chicago Press, 2002.

Dictionary of Russian Women Writers. Edited by Marina Ledkovsky, Charlotte Rosenthal, and Mary Zirin. Westport, CT: Greenwood Press, 1994.

Dixon, Laurinda S. *Perilous Chastity: Women and Illness in Pre-Enlightenment Art and Medicine.* Ithaca, NY: Cornell Universitiy Press, 1995.

Dolan, Frances, E. *Whores of Babylon: Catholicism, Gender and Seventeenth-Century Print Culture.* Ithaca, NY: Cornell University Press, 1999.

Donovan, Josephine. *Women and the Rise of the Novel, 1405–1726.* New York: St. Martin's Press, 1999.

Encyclopedia of Continental Women Writers. 2 vols. Ed. Katharina Wilson. New York: Garland, 1991.

De Erauso, Catalina. *Lieutenant Nun: Memoir of a Basque Transvestite in the New World.* Trans. Michele Ttepto and Gabriel Stepto; foreword by Marjorie Garber. Boston: Beacon Press, 1995.

Erdmann, Axel. *My Gracious Silence: Women in the Mirror of Sixteenth-Century Printing in Western Europe.* Luzern: Gilhofer and Rauschberg, 1999.

Erickson, Amy Louise. *Women and Property in Early Modern England.* London and New York: Routledge, 1993.

Ezell, Margaret J. M. *The Patriarch's Wife: Literary Evidence and the History of the Family.* Chapel Hill, NC: University of North Carolina Press, 1987.

———. *Social Authorship and the Advent of Print.* Baltimore: Johns Hopkins University Press, 1999.

———. *Writing Women's Literary History.* Baltimore: Johns Hopkins University Press, 1993.

Farrell, Michèle Longino. *Performing Motherhood: The Sévigné Correspondence.* Hanover, NH, and London: University Press of New England, 1991.

The Feminist Companion to Literature in English: Women Writers from the Middle Ages to the Present. Ed. Virginia Blain, Isobel Grundy, and Patricia Clements. New Haven, CT: Yale University Press, 1990.

The Feminist Encyclopedia of German Literature. Ed. Friederike Eigler and Susanne Kord. Westport, CT: Greenwood Press, 1997.

Feminist Encyclopedia of Italian Literature. Ed. Rinaldina Russell. Westport, CT: Greenwood Press, 1997.

Ferguson, Margaret W. *Dido's Daughters: Literacy, Gender, and Empire in Early Modern England and France.* Chicago: University of Chicago Press, 2003.

Ferguson, Margaret W., Maureen Quilligan, and Nancy J. Vickers, eds. *Rewriting the Renaissance: The Discourses of Sexual Difference in Early Modern Europe.* Chicago: University of Chicago Press, 1987.

Ferraro, Joanne M. *Marriage Wars in Late Renaissance Venice.* Oxford: Oxford University Press, 2001.

Fletcher, Anthony. *Gender, Sex and Subordination in England, 1500–1800.* New Haven, CT: Yale University Press, 1995.

French Women Writers: A Bio-Bibliographical Source Book. Ed. Eva Martin Sartori and Dorothy Wynne Zimmerman. Westport, CT: Greenwood Press, 1991.

Frye, Susan, and Karen Robertson, eds. *Maids and Mistresses, Cousins and Queens: Women's Alliances in Early Modern England.* Oxford: Oxford University Press, 1999.

Gallagher, Catherine. *Nobody's Story: The Vanishing Acts of Women Writers in the Marketplace, 1670–1820.* Berkeley: University of California Press, 1994.

Garrard, Mary D. *Artemisia Gentileschi: The Image of the Female Hero in Italian Baroque Art.* Princeton: Princeton University Press, 1989.

Gelbart, Nina Rattner. *The King's Midwife: A History and Mystery of Madame du Coudray.* Berkeley: University of California Press, 1998.

Glenn, Cheryl. *Rhetoric Retold: Regendering the Tradition from Antiquity through the Renaissance.* Carbondale and Edwardsville, IL: Southern Illinois University Press, 1997.

Goffen, Rona. *Titian's Women.* New Haven, CT: Yale University Press, 1997.

Goldberg, Jonathan. *Desiring Women Writing: English Renaissance Examples.* Stanford: Stanford University Press, 1997.

Goldsmith, Elizabeth C. *Exclusive Conversations: The Art of Interaction in Seventeenth-Century France.* Philadelphia: University of Pennsylvania Press, 1988.

———, ed. *Writing the Female Voice.* Boston: Northeastern University Press, 1989.

Goldsmith, Elizabeth C., and Dena Goodman, eds. *Going Public: Women and Publishing in Early Modern France.* Ithaca, NY: Cornell University Press, 1995.

Grafton, Anthony, and Lisa Jardine. *From Humanism to the Humanities: Education and the Liberal Arts in Fifteenth-and Sixteenth-Century Europe.* London: Duckworth, 1986.

Greer, Margaret Rich. *Maria de Zayas Tells Baroque Tales of Love and the Cruelty of Men.* University Park, PA: Pennsylvania State University Press, 2000.

Hackett, Helen. *Women and Romance Fiction in the English Renaissance.* Cambridge: Cambridge University Press, 2000.

Hall, Kim F. *Things of Darkness: Economies of Race and Gender in Early Modern England.* Ithaca, NY: Cornell University Press, 1995.

Hampton, Timothy. *Literature and the Nation in the Sixteenth Century: Inventing Renaissance France.* Ithaca, NY: Cornell University Press, 2001.

Hannay, Margaret, ed. *Silent but for the Word.* Kent, OH: Kent State University Press, 1985.

Hardwick, Julie. *The Practice of Patriarchy: Gender and the Politics of Household Authority in Early Modern France.* University Park, PA: Pennsylvania State University Press, 1998.

Harris, Barbara J. *English Aristocratic Women, 1450–1550: Marriage and Family, Property and Careers.* New York: Oxford University Press, 2002.

Harth, Erica. *Ideology and Culture in Seventeenth-Century France.* Ithaca, NY: Cornell University Press, 1983.

———. *Cartesian Women. Versions and Subversions of Rational Discourse in the Old Regime.* Ithaca, NY: Cornell University Press, 1992.

Harvey, Elizabeth D. *Ventriloquized Voices: Feminist Theory and English Renaissance Texts.* London and New York: Routledge, 1992.

Haselkorn, Anne M., and Betty Travitsky, eds. *The Renaissance Englishwoman in Print: Counterbalancing the Canon.* Amherst: University of Massachusetts Press, 1990.

Herlihy, David. "Did Women Have a Renaissance? A Reconsideration." *Medievalia et Humanistica* n.s. 13 (1985): 1–22.

Hill, Bridget. *The Republican Virago: The Life and Times of Catharine Macaulay, Historian.* New York: Oxford University Press, 1992.

A History of Central European Women's Writing. Ed. Celia Hawkesworth. New York: Palgrave Press, 2001.

A History of Women in the West.

Volume 1: *From Ancient Goddesses to Christian Saints.* Ed. Pauline Schmitt Pantel. Cambridge, MA: Harvard University Press, 1992.

Volume 2: *Silences of the Middle Ages.* Ed. Christiane Klapisch-Zuber. Cambridge, MA: Harvard University Press, 1992.

Volume 3: *Renaissance and Enlightenment Paradoxes.* Ed. Natalie Zemon Davis and Arlette Farge. Cambridge, MA: Harvard University Press, 1993.

A History of Women Philosophers. Ed. Mary Ellen Waithe. 3 vols. Dordrecht: Martinus Nijhoff, 1987.

A History of Women's Writing in France. Ed. Sonya Stephens. Cambridge: Cambridge University Press, 2000.

A History of Women's Writing in Germany, Austria and Switzerland. Ed. Jo Catling. Cambridge: Cambridge University Press, 2000.

A History of Women's Writing in Italy. Ed. Letizia Panizza and Sharon Wood. Cambridge: Cambridge University Press, 2000.

A History of Women's Writing in Russia. Ed. Alele Marie Barker and Jehanne M. Gheith. Cambridge: Cambridge University Press, 2002.

Hobby, Elaine. *Virtue of Necessity: English Women's Writing, 1646–1688.* London: Virago Press, 1988.

Horowitz, Maryanne Cline. "Aristotle and Women." *Journal of the History of Biology* 9 (1976): 183–213.

Howell, Martha. *The Marriage Exchange: Property, Social Place, and Gender in Cities of the Low Countries, 1300–1550.* Chicago: University of Chicago Press, 1998.

Hufton, Olwen H. *The Prospect before Her: A History of Women in Western Europe, 1: 1500–1800.* New York: HarperCollins, 1996.

Hull, Suzanne W. *Chaste, Silent, and Obedient: English Books for Women, 1475–1640.* San Marino, CA: Huntington Library, 1982.

Hunt, Lynn, ed. *The Invention of Pornography: Obscenity and the Origins of Modernity, 1500–1800.* New York: Zone Books, 1996.

Hutner, Heidi, ed. *Rereading Aphra Behn: History, Theory, and Criticism.* Charlottesville, VA: University Press of Virginia, 1993.

Hutson, Lorna, ed. *Feminism and Renaissance Studies.* New York: Oxford University Press, 1999.

Italian Women Writers: A Bio-Bibliographical Sourcebook. Ed. Rinaldina Russell. Westport, CT: Greenwood Press, 1994.

Jaffe, Irma B., with Gernando Colombardo. *Shining Eyes, Cruel Fortune: The Lives and Loves of Italian Renaissance Women Poets.* New York: Fordham University Press, 2002.

James, Susan E. *Kateryn Parr: The Making of a Queen.* Aldershot and Brookfield, UK: Ashgate Publishing Co., 1999.

Jankowski, Theodora A. *Women in Power in the Early Modern Drama.* Urbana, IL: University of Illinois Press, 1992.

Jansen, Katherine Ludwig. *The Making of the Magdalen: Preaching and Popular Devotion in the Later Middle Ages.* Princeton: Princeton University Press, 2000.

Jed, Stephanie H. *Chaste Thinking: The Rape of Lucretia and the Birth of Humanism.* Bloomington, IN: Indiana University Press, 1989.

Jordan, Constance. *Renaissance Feminism: Literary Texts and Political Models.* Ithaca, NY: Cornell University Press, 1990.

Kagan, Richard L. *Lucrecia's Dreams: Politics and Prophecy in Sixteenth-Century Spain.* Berkeley: University of California Press, 1990.

Kehler, Dorothea and Laurel Amtower, eds. *The Single Woman in Medieval and Early Modern England: Her Life and Representation.* Tempe, AZ: Medieval and Renaissance Texts and Studies, 2002.

Kelly, Joan. "Did Women Have a Renaissance?" In *Women, History, and Theory.* Chicago: University of Chicago Press, 1984. Also in *Becoming Visible: Women in European History,* Renate Bridenthal, Claudia Koonz, and Susan M. Stuard, eds. Third edition. Boston: Houghton Mifflin, 1998.

———. "Early Feminist Theory and the *Querelle des Femmes.*" In *Women, History, and Theory.*

Kelso, Ruth. *Doctrine for the Lady of the Renaissance.* Foreword by Katharine M. Rogers. Urbana, IL: University of Illinois Press, 1956, 1978.

King, Carole. *Renaissance Women Patrons: Wives and Widows in Italy, c. 1300–1550.* New York and Manchester: Manchester University Press (distributed in the U.S. by St. Martin's Press), 1998.

King, Margaret L. *Women of the Renaissance.* Foreword by Catharine R. Stimpson. Chicago: University of Chicago Press, 1991.

Krontiris, Tina. *Oppositional Voices: Women as Writers and Translators of Literature in the English Renaissance.* London and New York: Routledge, 1992.

Kuehn, Thomas. *Law, Family, and Women: Toward a Legal Anthropology of Renaissance Italy.* Chicago: University of Chicago Press, 1991.

Kunze, Bonnelyn Young. *Margaret Fell and the Rise of Quakerism.* Stanford: Stanford University Press, 1994.

Labalme, Patricia A., ed. *Beyond Their Sex: Learned Women of the European Past.* New York: New York University Press, 1980.

Laqueur, Thomas. *Making Sex: Body and Gender from the Greeks to Freud.* Cambridge, MA: Harvard University Press, 1990.

Larsen, Anne R., and Colette H. Winn, eds. *Renaissance Women Writers: French Texts/American Contexts.* Detroit, MI: Wayne State University Press, 1994.

Lerner, Gerda. *The Creation of Patriarchy* and *Creation of Feminist Consciousness, 1000–1870.* 2 vols. New York: Oxford University Press, 1986, 1994.

Levin, Carole, and Jeanie Watson, eds. *Ambiguous Realities: Women in the Middle Ages and Renaissance.* Detroit: Wayne State University Press, 1987.

Levin, Carole, et al. *Extraordinary Women of the Medieval and Renaissance World: A Biographical Dictionary.* Westport, CT: Greenwood Press, 2000.

Lewalsky, Barbara Kiefer. *Writing Women in Jacobean England.* Cambridge, MA: Harvard University Press, 1993.

Lewis, Jayne Elizabeth. *Mary Queen of Scots: Romance and Nation.* London: Routledge, 1998.

Lindsey, Karen. *Divorced Beheaded Survived: A Feminist Reinterpretation of the Wives of Henry VIII.* Reading, MA: Addison-Wesley Publishing Co., 1995.

Lochrie, Karma. *Margery Kempe and Translations of the Flesh.* Philadelphia: University of Pennsylvania Press, 1992.

Lougee, Carolyn C. *Le Paradis des Femmes: Women, Salons, and Social Stratification in Seventeenth-Century France.* Princeton: Princeton University Press, 1976.

Love, Harold. *The Culture and Commerce of Texts: Scribal Publication in Seventeenth-Century England.* Amherst, MA: University of Massachusetts Press, 1993.

MacCarthy, Bridget G. *The Female Pen: Women Writers and Novelists, 1621–1818.* Preface by Janet Todd. New York: New York University Press, 1994. Originally published by Cork University Press, 1946–47.

Maclean, Ian. *Woman Triumphant: Feminism in French Literature, 1610–1652.* Oxford: Clarendon Press, 1977.

———. *The Renaissance Notion of Woman: A Study of the Fortunes of Scholasticism and Medical Science in European Intellectual Life.* Cambridge: Cambridge University Press, 1980.

MacNeil, Anne. *Music and Women of the Commedia dell'Arte in the Late Sixteenth Century.* New York: Oxford University Press, 2003.

Maggi, Armando. *Uttering the Word: The Mystical Performances of Maria Maddalena de' Pazzi, a Renaissance Visionary.* Albany: State University of New York Press, 1998.

Marshall, Sherrin. *Women in Reformation and Counter-Reformation Europe: Public and Private Worlds.* Bloomington, IN: Indiana University Press, 1989.

Masten, Jeffrey. *Textual Intercourse: Collaboration, Authorship, and Sexualities in Renaissance Drama.* Cambridge: Cambridge University Press, 1997.

Matter, E. Ann, and John Coakley, eds. *Creative Women in Medieval and Early Modern Italy.* Philadelphia: University of Pennsylvania Press, 1994. Sequel to the Monson collection, below.

McLeod, Glenda. *Virtue and Venom: Catalogs of Women from Antiquity to the Renaissance.* Ann Arbor: University of Michigan Press, 1991.

Medwick, Cathleen. *Teresa of Avila: The Progress of a Soul.* New York: Alfred A. Knopf, 2000.

Meek, Christine, ed. *Women in Renaissance and Early Modern Europe.* Dublin-Portland: Four Courts Press, 2000.

Mendelson, Sara, and Patricia Crawford. *Women in Early Modern England, 1550–1720.* Oxford: Clarendon Press, 1998.

Merchant, Carolyn. *The Death of Nature: Women, Ecology and the Scientific Revolution.* New York: HarperCollins, 1980.

Merrim, Stephanie. *Early Modern Women's Writing and Sor Juana Inés de la Cruz.* Nashville, TN: Vanderbilt University Press, 1999.

Messbarger, Rebecca. *The Century of Women: The Representations of Women in Eighteenth-Century Italian Public Discourse.* Toronto: University of Toronto Press, 2002.

Miller, Nancy K. *The Heroine's Text: Readings in the French and English Novel, 1722–1782.* New York: Columbia University Press, 1980.

Miller, Naomi J. *Changing the Subject: Mary Wroth and Figurations of Gender in Early Modern England.* Lexington, KY: University Press of Kentucky, 1996.

Miller, Naomi J., and Gary Waller, eds. *Reading Mary Wroth: Representing Alternatives in Early Modern England.* Knoxville, TN: University of Tennessee Press, 1991.

Monson, Craig A., ed. *The Crannied Wall: Women, Religion, and the Arts in Early Modern Europe.* Ann Arbor: University of Michigan Press, 1992.

Musacchio, Jacqueline Marie. *The Art and Ritual of Childbirth in Renaissance Italy.* New Haven, CT: Yale University Press, 1999.

Newman, Barbara. *God and the Goddesses: Vision, Poetry, and Belief in the Middle Ages.* Philadelphia: University of Pennsylvania Press, 2003.

Newman, Karen. *Fashioning Femininity and English Renaissance Drama.* Chicago and London: University of Chicago Press, 1991.

Okin, Susan Moller. *Women in Western Political Thought.* Princeton: Princeton University Press, 1979.

Ozment, Steven. *The Bürgermeister's Daughter: Scandal in a Sixteenth-Century German Town.* New York: St. Martin's Press, 1995.

Pacheco, Anita, ed. *Early [English] Women Writers, 1600–1720.* New York and London: Longman, 1998.

Pagels, Elaine. *Adam, Eve, and the Serpent.* New York: HarperCollins, 1988.

Panizza, Letizia, ed. *Women in Italian Renaissance Culture and Society.* Oxford: European Humanities Research Centre, 2000.

Parker, Patricia. *Literary Fat Ladies: Rhetoric, Gender and Property.* London and New York: Methuen, 1987.

Pernoud, Regine, and Marie-Veronique Clin. *Joan of Arc: Her Story.* Rev. and trans. Jeremy DuQuesnay Adams. New York: St. Martin's Press, 1998. French original, 1986.

Perry, Mary Elizabeth. *Crime and Society in Early Modern Seville.* Hanover, NH: University Press of New England, 1980.

———. *Gender and Disorder in Early Modern Seville.* Princeton: Princeton University Press, 1990.

Petroff, Elizabeth Alvilda, ed. *Medieval Women's Visionary Literature.* New York: Oxford University Press, 1986.

Perry, Ruth. *The Celebrated Mary Astell: An Early English Feminist.* Chicago: University of Chicago Press, 1986.

Rabil, Albert. *Laura Cereta: Quattrocento Humanist.* Binghamton, NY: Medieval and Renaissance Texts and Studies, 1981.

Ranft, Patricia. *Women in Western Intellectual Culture, 600–1500.* New York: Palgrave, 2002.

Rapley, Elizabeth. *A Social History of the Cloister: Daily Life in the Teaching Monasteries of the Old Regime.* Montreal: McGill-Queen's University Press, 2001.

Raven, James, Helen Small, and Naomi Tadmor, eds. *The Practice and Representation of Reading in England.* Cambridge: Cambridge University Press, 1996.

Reardon, Colleen. *Holy Concord within Sacred Walls: Nuns and Music in Siena, 1575–1700.* Oxford: Oxford University Press, 2001.

Reiss, Sheryl E., and David G. Wilkins, ed. *Beyond Isabella: Secular Women Patrons of Art in Renaissance Italy.* Kirksville, MO: Truman State University Press, 2001.

Rheubottom, David. *Age, Marriage, and Politics in Fifteenth-Century Ragusa.* Oxford: Oxford University Press, 2000.

Richardson, Brian. *Printing, Writers and Readers in Renaissance Italy.* Cambridge: Cambridge University Press, 1999.

Riddle, John M. *Contraception and Abortion from the Ancient World to the Renaissance.* Cambridge, MA: Harvard University Press, 1992.

———. *Eve's Herbs: A History of Contraception and Abortion in the West.* Cambridge, MA: Harvard University Press, 1997.

Rose, Mary Beth. *The Expense of Spirit: Love and Sexuality in English Renaissance Drama.* Ithaca, NY: Cornell University Press, 1988.

———. *Gender and Heroism in Early Modern English Literature.* Chicago: University of Chicago Press, 2002.

———, ed. *Women in the Middle Ages and the Renaissance: Literary and Historical Perspectives.* Syracuse: Syracuse University Press, 1986.

Rosenthal, Margaret F. *The Honest Courtesan: Veronica Franco, Citizen and Writer in Sixteenth-Century Venice.* Foreword by Catharine R. Stimpson. Chicago: University of Chicago Press, 1992.

Sackville-West, Vita. *Daughter of France: The Life of La Grande Mademoiselle.* Garden City, NY: Doubleday, 1959.

Sánchez, Magdalena S. *The Empress, the Queen, and the Nun: Women and Power at the Court of Philip III of Spain.* Baltimore: Johns Hopkins University Press, 1998.

Schiebinger, Londa. *The Mind Has No Sex?: Women in the Origins of Modern Science.* Cambridge, MA: Harvard University Press, 1991.

———. *Nature's Body: Gender in the Making of Modern Science.* Boston: Beacon Press, 1993.

Schutte, Anne Jacobson, Thomas Kuehn, and Silvana Seidel Menchi, eds. *Time, Space, and Women's Lives in Early Modern Europe.* Kirksville, MO: Truman State University Press, 2001.

Schofield, Mary Anne, and Cecilia Macheski, eds. *Fetter'd or Free? British Women Novelists, 1670–1815.* Athens, OH: Ohio University Press, 1986.

Shannon, Laurie. *Sovereign Amity: Figures of Friendship in Shakespearean Contexts.* Chicago: University of Chicago Press, 2002.

Shemek, Deanna. *Ladies Errant: Wayward Women and Social Order in Early Modern Italy.* Durham, NC: Duke University Press, 1998.

Smith, Hilda L. *Reason's Disciples: Seventeenth-Century English Feminists.* Urbana, IL: University of Illinois Press, 1982.

———. *Women Writers and the Early Modern British Political Tradition.* Cambridge: Cambridge University Press, 1998.

Sobel, Dava. *Galileo's Daughter: A Historical Memoir of Science, Faith, and Love.* New York: Penguin Books, 2000.

Sommerville, Margaret R. *Sex and Subjection: Attitudes to Women in Early-Modern Society.* London: Arnold, 1995.

Soufas, Teresa Scott. *Dramas of Distinction: A Study of Plays by Golden Age Women.* Lexington, KY: The University Press of Kentucky, 1997.

Spencer, Jane. *The Rise of the Woman Novelist: From Aphra Behn to Jane Austen.* Oxford: Basil Blackwell, 1986.

Spender, Dale. *Mothers of the Novel: 100 Good Women Writers before Jane Austen.* London and New York: Routledge, 1986.

Sperling, Jutta Gisela. *Convents and the Body Politic in Late Renaissance Venice.* Foreword by Catharine R. Stimpson. Chicago: University of Chicago Press, 1999.

Steinbrügge, Lieselotte. *The Moral Sex: Woman's Nature in the French Enlightenment.* Trans. Pamela E. Selwyn. New York: Oxford University Press, 1995.

Stocker, Margarita. *Judith, Sexual Warrior: Women and Power in Western Culture.* New Haven, CT: Yale University Press, 1998.

Stretton, Timothy. *Women Waging Law in Elizabethan England.* Cambridge: Cambridge University Press, 1998.

Stuard, Susan M. "The Dominion of Gender: Women's Fortunes in the High Middle Ages." In *Becoming Visible: Women in European History,* ed. Renate Bridenthal, Claudia Koonz, and Susan M. Stuard. Third edition. Boston: Houghton Mifflin, 1998.

Summit, Jennifer. *Lost Property: The Woman Writer and English Literary History, 1380–1589.* Chicago: University of Chicago Press, 2000.

Surtz, Ronald E. *The Guitar of God: Gender, Power, and Authority in the Visionary World of Mother Juana de la Cruz (1481–1534).* Philadelphia: University of Pennsylvania Press, 1991.

———. *Writing Women in Late Medieval and Early Modern Spain: The Mothers of Saint Teresa of Avila.* Philadelphia: University of Pennsylvania Press, 1995.

Teague, Frances. *Bathsua Makin, Woman of Learning.* Lewisburg, PA: Bucknell University Press, 1999.

Todd, Janet. *The Secret Life of Aphra Behn.* London, New York, and Sydney: Pandora, 2000.

———. *The Sign of Angelica: Women, Writing and Fiction, 1660–1800.* New York: Columbia University Press, 1989.

Valenze, Deborah. *The First Industrial Woman.* New York: Oxford University Press, 1995.

Van Dijk, Susan, Lia van Gemert, and Sheila Ottway, eds. *Writing the History of Women's Writing: Toward an International Approach.* Proceedings of the Colloquium, Amsterdam, 9–11 September. Amsterdam: Royal Netherlands Academy of Arts and Sciences, 2001.

Vickery, Amanda. *The Gentleman's Daughter: Women's Lives in Georgian England.* New Haven, CT: Yale University Press, 1998.

Vollendorf, Lisa, ed. *Recovering Spain's Feminist Tradition.* New York: Modern Language Association, 2001.

Walker, Claire. *Gender and Politics in Early Modern Europe: English Convents in France and the Low Countries.* New York: Palgrave, 2003.

Wall, Wendy. *The Imprint of Gender: Authorship and Publication in the English Renaissance.* Ithaca, NY: Cornell University Press, 1993.

Walsh, William T. *St. Teresa of Avila: A Biography.* Rockford, IL: TAN Books and Publications, 1987.

Warner, Marina. *Alone of All Her Sex: The Myth and Cult of the Virgin Mary.* New York: Knopf, 1976.

Warnicke, Retha M. *The Marrying of Anne of Cleves: Royal Protocol in Tudor England.* Cambridge: Cambridge University Press, 2000.

Watt, Diane. *Secretaries of God: Women Prophets in Late Medieval and Early Modern England.* Cambridge, UK: D.S. Brewer, 1997.

Weber, Alison. *Teresa of Avila and the Rhetoric of Femininity.* Princeton: Princeton University Press, 1990.

Welles, Marcia L. *Persephone's Girdle: Narratives of Rape in Seventeenth-Century Spanish Literature.* Nashville: Vanderbilt University Press, 2000.

Whitehead, Barbara J., ed. *Women's Education in Early Modern Europe: A History, 1500–1800.* New York and London: Garland Publishing Co., 1999.

Wiesner, Merry E. *Women and Gender in Early Modern Europe.* Cambridge: Cambridge University Press, 1993.

———. *Working Women in Renaissance Germany.* New Brunswick, NJ: Rutgers University Press, 1986.

Willard, Charity Cannon. *Christine de Pizan: Her Life and Works.* New York: Persea Books, 1984.

Winn, Colette, and Donna Kuizenga, eds. *Women Writers in Pre-Revolutionary France.* New York: Garland Publishing, 1997.

Woodbridge, Linda. *Women and the English Renaissance: Literature and the Nature of Womankind, 1540–1620.* Urbana, IL: University of Illinois Press, 1984.

Woods, Susanne. *Lanyer: A Renaissance Woman Poet.* New York: Oxford University Press, 1999.

Woods, Susanne, and Margaret P. Hannay, eds. *Teaching Tudor and Stuart Women Writers.* New York: Modern Language Association, 2000.

INDEX